EAI/Springer Innovations in Communication and Computing

Series Editor
Imrich Chlamtac, European Alliance for Innovation, Ghent, Belgium

The impact of information technologies is creating a new world yet not fully understood. The extent and speed of economic, life style and social changes already perceived in everyday life is hard to estimate without understanding the technological driving forces behind it. This series presents contributed volumes featuring the latest research and development in the various information engineering technologies that play a key role in this process. The range of topics, focusing primarily on communications and computing engineering include, but are not limited to, wireless networks; mobile communication; design and learning; gaming; interaction; e-health and pervasive healthcare; energy management; smart grids; internet of things; cognitive radio networks; computation; cloud computing; ubiquitous connectivity, and in mode general smart living, smart cities, Internet of Things and more. The series publishes a combination of expanded papers selected from hosted and sponsored European Alliance for Innovation (EAI) conferences that present cutting edge, global research as well as provide new perspectives on traditional related engineering fields. This content, complemented with open calls for contribution of book titles and individual chapters, together maintain Springer's and EAI's high standards of academic excellence. The audience for the books consists of researchers, industry professionals, advanced level students as well as practitioners in related fields of activity include information and communication specialists, security experts, economists, urban planners, doctors, and in general representatives in all those walks of life affected ad contributing to the information revolution.
Indexing: This series is indexed in Scopus, Ei Compendex, and zbMATH.

About EAI - EAI is a grassroots member organization initiated through cooperation between businesses, public, private and government organizations to address the global challenges of Europe's future competitiveness and link the European Research community with its counterparts around the globe. EAI reaches out to hundreds of thousands of individual subscribers on all continents and collaborates with an institutional member base including Fortune 500 companies, government organizations, and educational institutions, provide a free research and innovation platform. Through its open free membership model EAI promotes a new research and innovation culture based on collaboration, connectivity and recognition of excellence by community.

Manju Khari • Deepti Bala Mishra
Biswaranjan Acharya • Ruben Gonzalez Crespo
Editors

Optimization of Automated Software Testing Using Meta-Heuristic Techniques

Editors
Manju Khari
School of Computer & Systems Sciences
Jawaharlal Nehru University
New Delhi, Delhi, India

Deepti Bala Mishra
Department of MCA
GITA Autonomous College
Bhubaneswar, India

Biswaranjan Acharya
Department of Computer Engineering-AI
Marwadi University
Rajkot, Gujarat, India

Ruben Gonzalez Crespo
Computer Science and Technology
Universidad Internacional de La Rioja
La Rioja, Spain

ISSN 2522-8595 ISSN 2522-8609 (electronic)
EAI/Springer Innovations in Communication and Computing
ISBN 978-3-031-07296-3 ISBN 978-3-031-07297-0 (eBook)
https://doi.org/10.1007/978-3-031-07297-0

This Springer imprint is published by the registered company Springer Nature Switzerland AG
The registered company address is: Gewerbestrasse 11, 6330 Cham, Switzerland

Preface

Test automation is now ubiquitous, and almost assumed in large segments of the research. Agile processes and test-driven development are now widely known and used for implementation and deployment. This book presents software testing as a practical engineering activity, essential to producing high-quality software. This book is beneficial for an undergraduate or graduate course on software testing and software engineering, and as a resource for software test engineers and developers. This book has a number of unique features:

1. It includes a landscape of test coverage criteria with a novel and extremely simple structure. At a technical level, software testing is based on satisfying coverage criteria. The book's central observation is that there are few truly different coverage criteria, each of which fits easily into one of four categories: graphs, logical expressions, input space, and syntax structures.
2. It projects a balance of theory and practical application, presenting testing as a collection of objective, quantitative activities that can be measured and repeated. The theoretical concepts are presented when needed to support the practical activities that researchers and test engineers follow.
3. It assumes the reader is learning to be a researcher whose goal is to produce the best possible software with the lowest possible cost. The concepts in this book are well grounded in theory, are practical, and most are currently in use.

Through this book an effort to in support of the idea of promoting software testing and establishing as to software testing is made possible.

Chapter 1

In Chap. 1, test suite minimization is done with an intention of optimizing the test suite, and software faults detection and localization as well as adjoining activities are focused on. In this chapter, code coverage and mutant algorithms are used to generate the compact test cases on which an algorithm is applied for identifying and locating errors. To optimize the test cases, NSGA-II algorithm is used. Defects4j repository has been used for generating and performing tests.

Chapter 2

Chapter 2 focuses on mutation testing, which is the fault-based software testing approach that is widely applicable for assessing the effectiveness of a test suite. The test suite effectiveness is measured through artificial seeding of faults into the programs under test. Six open-source mutation testing tools and JAVA-based MTT (Jester, Javamut, MuJava, Jumble, Judy, and Javalanche) are analyzed. The results are based on the performance of various JAVA programs and two real-life applications. Benchmark comparison among the MTT is presented in terms of mutants, mutation operator, mutation score, and quality output.

Chapter 3

In Chap. 3, the authors present MBT and state-based test case generation using a state chart diagram. Firstly, the authors review the main concepts and techniques in MBT. Then, in the next step, they review the most common modeling formalisms for state chart diagram, with focus on various state-based coverage criteria. Subsequently, the authors propose methods for a synchronous state-based testing approach to generate test cases.

Chapter 4

In Chap. 4, the Author designed and developed a TCP technique to enhance the fault detection rate of test cases at the early execution of the test suite. The developed algorithm was examined with two benchmark algorithms on four subject programs to evaluate the performance of the algorithm. APFD metrics are used as performance evaluation metrics and the performance of the developed algorithm outperforms both of the benchmark algorithms.

Chapter 5

In Chap. 5, authors analyze the already available and enhanced testing techniques for the improved and good quality product. Some recent research studies have been summed up in this work as software testing is acquiring more significance these days.

Chapter 6

In Chap. 6, authors identify and analyze an existing research paper previously conducted by different researchers on predicting software reliability using a machine learning approach in the context of formulated research questions.

Chapter 7

In Chap. 7, a systematic approach to finding bugs means errors or different other defects in a running application which are ready to tested. It also helps to analyze the actual programs and to lower the cost of finding errors. And different EAs like GA-, PSO-, ACO-, and ABCO-based methods have been already proposed to generate the optimized test cases.

Chapter 8

Chapter 8 represents a use case of optimization of software testing in different wireless sensor network applications. The survey in the paper also shows that the use of a metaheuristic is not limited to WSN, and the use of a metaheuristic in automated software testing is exemplary. In the field of software testing, optimization of test cases and increasing usability are a few tasks that can be optimized with the help of metaheuristic algorithms.

Chapter 9

In Chap. 9, the author develops my CHIP-8 emulator for software testing strategy for playing online games on many platforms. The author lists each instruction explaining what it does and how it carries out it while providing the detailed documentation of our CHIP-8 emulator and thus, providing metaheuristic high-level solutions to fix them.

Chapter 10

Chapter 10 describes defects maintainability prediction of the software. This chapter evaluates the mentioned scenario by using maintainability index and defect data. The maintainability index is computed using the object-oriented metrics of the software.

Chapter 11

The book ends with Chap. 11, which develops a hybrid metaheuristic encryption approach employing software testing for secure data transmission named EncryptoX. The main objective behind doing this project report is to gain skills and knowledge regarding various cryptography and storage techniques used in software testing.

New Delhi, India — Manju Khari
Bhubaneswar, India — Deepti Bala Mishra
Rajkot, Gujarat, India — Biswaranjan Acharya
La Rioja, Spain — Ruben Gonzalez Crespo

Contents

NGA-II-Based Test Suite Minimization in Software

Renu Dalal, Manju Khari, Tushar Singh Bhal, and Kunal Sharma

1 Introduction

Developing software is one of the major works that is being done in the industry in this era of technology. For developing software one of the major tasks is testing for errors and issues. Running the complete test suite without minimizing is a tedious job as it will induce a big load on the system and the operation under execution. Test case minimization is one of the options to reduce the test suite. For testing purposes, the first step is to create test suites in which some operations are defined or a set of information in which the software has to perform and the results of which describe the ability of the software to perform under that kind of task.

But after creation of a test suite, the next step is to minimize the test suite as it can contain many redundant and faulty tests which have to be removed to improve the efficiency of the testing. The load on the machine gets reduced if proper minimization is done. Test case minimization is used to getting the compacted test case. This aids in testing that modification done in software program has not affected the unmodified part of the software. Identifying and locating errors is one of the major tasks which has to be done to minimize the test suite. But performing them at the same time is a different issue as they are subsequent activities. At first, fault detection in the test suite using failing test cases is done, and then localization is done by using the pass and fail information of the test suit.

R. Dalal (✉)
Department of Information Technology, MSIT, GGSIP University, Delhi, India

M. Khari
School of Computer and Systems Sciences, Jawaharlal Nehru University, Delhi, India

T. S. Bhal · K. Sharma
Department of Computer Science, AIACT & R, GGSIP University, Delhi, India

M. Khari et al. (eds.), *Optimization of Automated Software Testing Using Meta-Heuristic Techniques*, EAI/Springer Innovations in Communication and Computing, https://doi.org/10.1007/978-3-031-07297-0_1

Software testing is one of the major processes in all the SDLC phases present. Testing takes a lot of time and resources present at our end. Research says that in about 50% of the total time given in development, 25% of it should be given only to debugging. Test suite minimization helps in this process by reducing the time and computing power applied on the tasks. As both detecting and localizing faults should be performed one after another if we can combine these processes, this can significantly reduce the work and load applied. The aim of this chapter is to achieve greater efficiency as possible in minimizing and to achieve better CPU utilization, memory utilization, and disk usage.

A software always gets updated, and new functionality is added from time to time due to which new test cases are added in the test suite. After some updates, there are some test cases which are no longer needed and are an overhead over the system, so they have to be removed or should not be considered that is why it is a must to perform minimization over the test suite. It also becomes easy to detect faults in the minimized suite. As mentioned above, the detection and localization tasks are done one after another. Combining them is a difficult process but essential as performing the in concurrently takes more time and usage of other resources. For combining them offline techniques can be used. The reduced test suite obtained after these processes can be used for regression testing. Vidács et al. and many other researchers work on this approach. They proposed an approach for combining both these processes and minimizing the test suite.

For doing all this, multi-objective optimization algorithms come into play. Multi-objective optimization algorithm deals in the domain of optimization problems where multiple objective functions are optimized. The solution obtained is known as non-dominated, Pareto adequate or non-inferior, and Pareto optimal, if none of the objective functions can enhance the value without deteriorating the other objective values. An assessment is done in this study on projects taken from Defects4J repository [1–3]. This assessment is performed by using NSGA-II, coverage, and mutation algorithms. The main reason for undertaking minimization is:

- To reduce time for testing purpose
- Less system requirement
- Redundant test case elimination
- To predict faults easily

2 Background

2.1 What Is Test Suite?

Test suite is a collection of tests which helps testers in executing and performing testing and reporting faults and errors present in the software. A single test case can be added to many test suites. A test suite consists of many test cases which describe the various conditions which the software has to encounter while being operated. It can also be defined as a collection of scenarios which define the scope of testing for a given execution environment.

Test suites are used for identifying gaps in testing efforts where successful completion of a test case can occur before the next step begins. Test suites are also useful like build verification test, smoke test, end-to-end integration test, and functional verification test.

A test suite can divide into three types:

- *Static Suite*: In this suite, creation of a collection of named scenarios will remain static once defined. In this type sequence of execution is always guaranteed. The order of execution is defined in this type of suite.
- *Filter-Based Suite*: In this suite, custom sets of fields are defined using filter parameters. This type of suite is used for targeted testing of specified portions.
- *Requirement-Based Suite*: Here test suite is created based on the user requirements. This is mostly used in agile environments.

2.2 *Minimization of Test Suite*

A test suite contains a huge amount of test cases, and executing them is an annoying task. Many researches have been done to minimize this annoying task. A test suite is minimized by removing the redundant test cases and removing those cases which are no longer needed, or the functionality has been removed in the updated version of the software [4].

2.3 *Partitioning*

Partitioning can be divided into classes into equivalent partitions. This idea is based on the concept of equivalence partition in set theory. These partitions are undistinguishable. Majorly two types of partitions are accessible: statement partitioning and mutation partitioning. Statement partitioning is based on the code coverage. It means portioning the code segment on the basis of their code coverage information into diverse classes. Mutant partitioning is another type of partitioning method in which the mutants are partitioned on the basis of their kill information by test cases. Mutants can be partitioned into various partitions, for measuring this a factor d score is introduced by the researchers, it provides the information of number of mutants differentiate by the considered test-cases [5].

2.4 *Optimization Algorithms*

Optimization algorithms are used for optimizing test cases. These algorithms are used to reduce the test suite size and to produce results which can be used further in multi-objective tasks. There are many optimization algorithms present. It can broadly be classified into two categories, that is, single objective and multi-objective [6]. Multi-objective includes NSGA-II, NSGA-III, MO-PSO, MO-BAT, etc.

3 Defects4J

3.1 *About Defects4J Repository*

Defects4J is a repository which contains a collection of reproducible bugs which is used for advancing software research. Defects4J consists of many projects, and there are many versions of each project which can be used to generate test suites of various types. It contains 835 bugs from many open-source projects like Chart, Cli, Closure, Math, Time, Csv, etc. Test suites are generated by using some generator functions and then providing a version of the project. It also provides the support for integrating any methods of applying algorithms outside the repository scope. It comes with the support of applying basic coverage and mutation algorithms [7]. Defects4J comes with basic functionality like performing checkout and compiling and performing testing on a test case.

4 Code Coverage

The code coverage is a metric that illustrates the extent of the source program code that has been tested. It is a part of white box testing. It is used to determine the quantitative measure of the code. It generates the result of the test suite's code coverage. There are many reasons why we use code coverage, and some of them are to: (1) Offers Quantitative measurements. (2) Describes the extent to which the code is tested. (3) Calculate test implementation efficiency [8–10]. There are many techniques in which code coverage can be performed such as:

- Decision coverage
- Statement coverage
- Toggle coverage
- Branch coverage
- FSM coverage

In this chapter statement and branch coverage are used.

4.1 Statement Coverage

Statement coverage involves execution of all the executable statements in the source code at least once. It is white box testing technique. It is used to calculate the number of statements which can be executed on the given requirements. Here as a part of white box testing, the aim is to understand the working of the code at internal levels. Its main goal is to include all the finite routes, lines, and statements present in the source code. Maximization of statement coverage intends to discover the minimized test case and can enhance its value. The statement coverage metric is defined as Eq. 1:

$$SC = \frac{\left|\{s \in M \mid s \text{ covered by } T\}\right|}{|M|} \tag{1}$$

Here, $|M|$ means total number of statements present in the source program.

4.2 Branch Coverage

The outcome of the code module is tested in branch coverage. Branch coverage's main goal is to make sure every possible branch is tested. It tells us about the independent code segments present in the codes [11]. The branch coverage ensures that every section of each control structure may be examined at least once.

Maximization of branch coverage helps in finding the minimized test case, and it maximizes the value of branch coverage. The metric of branch coverage is represented in Eq. 2.

$$\text{Branchcover} = \frac{\left|\{b \in P \mid b \text{ covered by } T\}\right|}{|P|} \tag{2}$$

Here, $|P|$ represents the total number of branches in the code.

5 Proposed Approach

5.1 Workflow of Approach

Test suite minimization is required in software for the same the proposed approached in represented in the Fig. 1

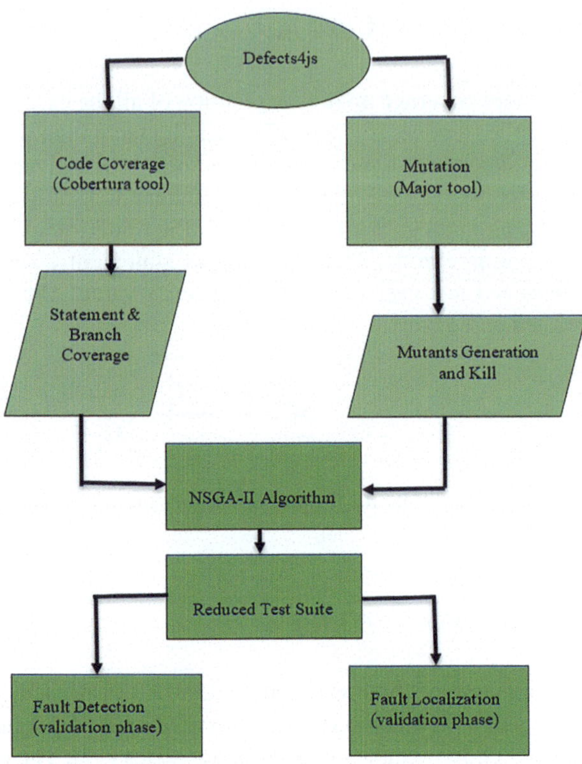

Fig. 1 Workflow of proposed approach

5.2 Optimization NSGA-II Algorithm

NSGA II is the multi-objective algorithm which comes in the class of optimization. It stands for elitist non-dominated sorting genetic algorithm. This algorithm is both elitism preserving and diversity preserving. Elitist means it keeps the best solution for the next iteration from the current one. Non-dominated searching means if there are two individuals A and B, A dominates to B, if and only if there is no objective of A worse than that objective of B and there is at least one objective of A better than that objective of B. The objective of non-dominated sorting is to find out which individual belongs to which front. Mathematically domination is:

A(x_1, y_1) dominates B(x_2, y_2) when : (x_1<=x_2 and y_1<=y_2) and (x_1<x_2 or y_1<y_2)

One of the fronts may not fit properly in the size of the parent population as before for this *crowding distance* is used. To keep a good spread in NSGA-II and avoid local maxima or minima, crowding distance decides which individuals are added to the new population. Individuals with higher crowding distance are picked first. After this new offspring is created which has the same size as the parent. This process happens in three phases tournament selection, crossover, and finally mutation. All this happens for some iterations, and then the result is taken.

Pseudo Code of the Algorithm

Fast sNon-Dominated Sort:

```
for every p ∈P
    S_p = Φ
    n = 0
    for every q∈P
        if q<p then
            S_p = S_p U{q}
        else if q<p then
            n = n +1
    if n = 0 then
        p_rank = 1
        F_1 = F_1U{q}
i = 1
while F_i ≠Φ
    Q = Φ
    for every q∈F_i
        for every q∈S_p
            n_q = n_q -1
            if n_q = 0
                q_rank = i+1
                Q = QU{q}
    i = i + 1
    F_i = Q
```

Crowding_Distance_Assignment

```
l = | I |
for every i, set I[i] dist = 0
for every objective m
    I = sort(I, m)
    I[1]dist = I[1]dist = ∞
    for i = 2 to (l-1)
                I[i]dist = I[i]dist + (I[i+1].m - I[i-1].m)/
(fmax_m    fmin_m)
```

Final_Step

```
R_t = P_tUQ__t
F = Fast_Non_Dominated_Sort(Rt)
```

Table 1 Project and version chosen with number of bugs present in each

Identifier	Project-name	No. of bugs	Active bugs ID
Lang	Commons-lang	64	1,3–65
Closure	Closure-compiler	174	1-62,64-92,94-0
Time	Joda-time	26	1–20, 22–27

```
P_t+1 = Φ and i = 1
until | P_t+1 | + | F_i | ≤ N

Crowding_Distance_Assignment(Fi)

    P_t+1 = P_t+1UF_i
    i = i+1
Sort(F_i, <n)
P_t+1 = P_t+1UF_i[1: (N- | P_t+1|)]
Q_t+1 = make-new-pop(P_t+1)
 t = t+1
```

5.3 Performing Coverage and Mutation

There are many projects and versions present in the Defects4J repository. This table shows the details of the work with their versions, number of bugs present in them, and active bugs ID. Table 1 shows projects used from the Defects4J repository.

Coverage task was performed on a shell in Linux. First, set a path of Defects4J in the terminal so that the system can understand where to look for commands. Now take the created test suite and checkout using the checkout command. This creates an executable folder where the data goes and stays for execution. After that perform coverage from the function coverage. This will perform coverage on the selected project. At the end it gives us the number of statements it covered and removes those redundant cases.

6 Results and Analysis

6.1 Result Obtained

After performing the experiments, the test suite was minimized. The images above show the reduction in the test suite as the number of test cases which were being covered earlier was more than being covered now by the same algorithm. It means that the redundant test cases were removed from the suite. It can be seen properly in each project taken by this work. This means that the approach taken was a successful one. Projects are Lang Project represented in Figs. 2 and 3; then Time project results represented in Figs. 4 and 5; and closure project results depicts in Figs 6 and 7.

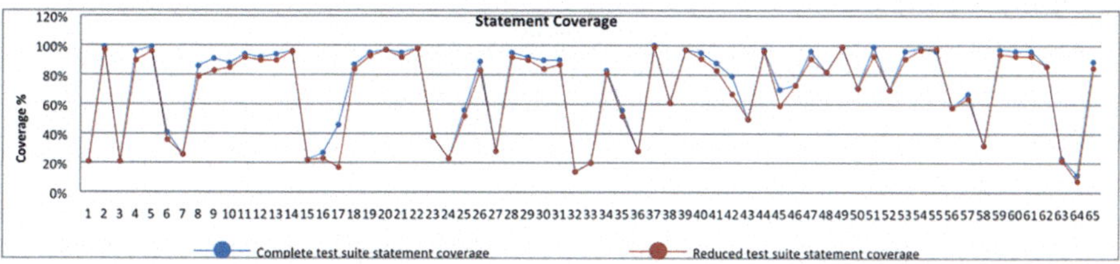

Fig. 2 Statement coverage for the Lang project

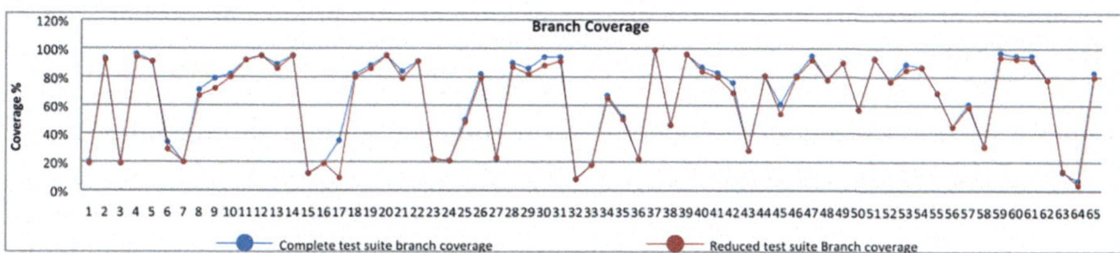

Fig. 3 Branch coverage for the Lang project

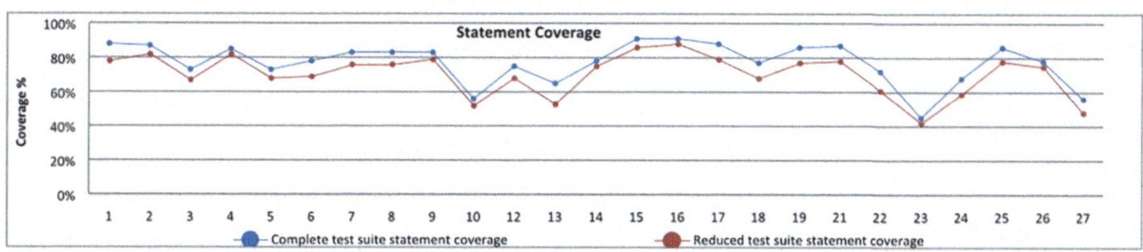

Fig. 4 Statement coverage for the time project

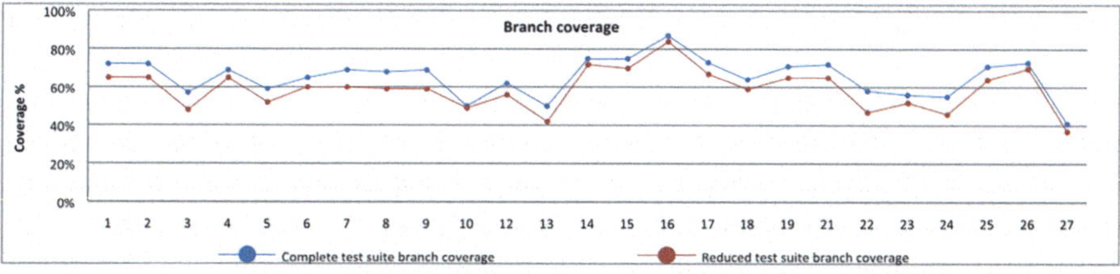

Fig. 5 Branch coverage for the time project

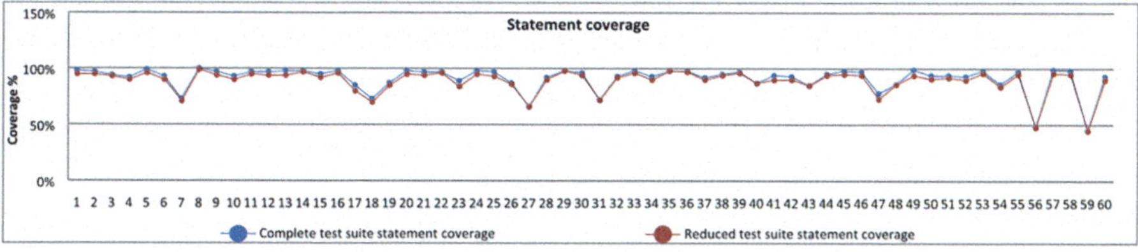

Fig. 6 Statement coverage for the closure project

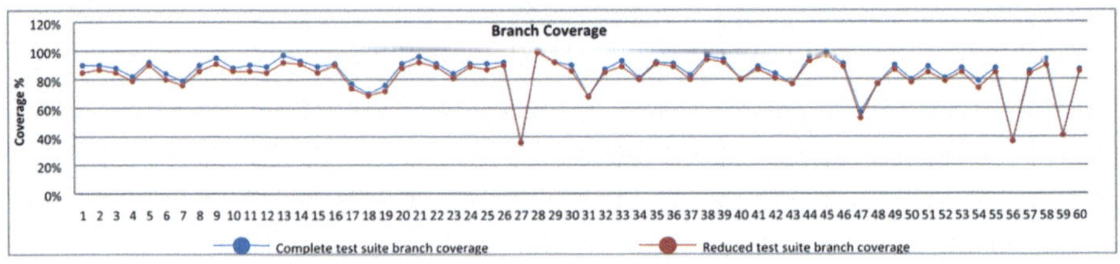

Fig. 7 Branch coverage for the closure project

6.1.1 Graphs

Graphs represent the amount of coverage percentage after applying the algorithms. X-axis of the graph represents the test suite related to a specific bug, and Y-axis represents the coverage percentage. Blue dots represent complete test suite coverage, and red dots represent coverage after reducing the suite. For each project there are two graphs, one for statement coverage and the other for branch coverage.

Lang Project

Time Project

Closure Project

6.2 Errors Occurred

While performing mutation on the projects, it was encountered that the mutants generated were not yielding the results as we expected. The mutants which could have given us better results were killed while mutation testing. The Defects4J repository has a built-in operation for creating mutants and performing mutation testing, and it was tried, but didn't work and gave some error. To resolve this, select the issues section of the repository, and it was found that there was this problem still unsolved. Then creating some mutants in the project still didn't work. Due to all these errors, the second set which was supposed to pass through the NSGA-II algorithm was not generated. So, it passed the only generated suite through the algorithm, by dividing it into two separate halves. The results were unexpectedly low. Due to this problem, the coverage part was performed, and the final validation step was skipped. Still the results were not that bad. The test suite was reduced to some extent.

7 Conclusion and Future Work

Software which have a lot of functionalities require a big test suite for testing purposes. So, test suite minimization is done. And with this minimization it becomes easy for a tester to identify and locate software faults present in the code. Here the purpose was to minimize that test suite and correctly detect and localize the faults present in the software. The approach included coverage, mutation, and then performing optimization using genetic algorithms. The average percent of test suite reduction achieved was only 62% as one of the parts didn't work properly as expected, if it would have then it was expected to achieve an accuracy of 72%. In future work, it is necessary to implement the mutation testing part which could have improved the accuracy further. So, it can implement that functionality for increasing the scores further. With developing technology and new nature-inspired algorithms coming in picture, this domain is getting more and more scope of enhancement.

References

1. Deb, K., Pratap, A., Agarwal, S., & Meyarivan, T. A. M. T. (2002). A fast and elitist multiobjective genetic algorithm: NSGA-II. *IEEE Transactions on Evolutionary Computation, 6*(2), 182–197.
2. Khari, M., Dalal, R., & Rohilla, P. (2020). Extended paradigms for botnets with WoT applications: A review. *Smart Innovation of Web of Things*, 105.
3. Khari, M., Dalal, R., Misra, U., & Kumar, A. (2020). AndroSet: An automated tool to create datasets for android malware detection and functioning with WoT. *Smart Innovation of Web of Things*, 187.
4. Correia, D., Abreu, R., Santos, P., & Nadkarni, J. (2019). MOTSD: A multi-objective test selection tool using test suite diagnosability. In *Proceedings of the 2019 27th ACM joint meeting on European software engineering conference and symposium on the foundations of software engineering* (pp. 1070–1074).
5. Just, R. (2014). The Major mutation framework: Efficient and scalable mutation analysis for Java. In *Proceedings of the 2014 international symposium on software testing and analysis* (pp. 433–436). ACM.
6. Nature Inspired Algorithm: Nature-Inspired Metaheuristic Algorithms by Xin-She Yang. https://www.researchgate.net/publication/235979455_Nature-Inspired_Metaheuristic_Algorithms
7. Defects4j. https://github.com/rjust/defects4j
8. Cobertura. http://cobertura.sourceforge.net/
9. Mutation tool MAJOR. https://mutation-testing.org/doc/
10. Code Coverage. https://www.guru99.com/code-coverage.html
11. Mutation testing. https://www.guru99.com/mutation-testing.html

Comparison and Validation of Mutation Testing Tools Based on Java Language

Manju Khari

1 Introduction

Mutation testing is the testing approach that introduces errors by making simple artificial modifications into source code. The introduction of artificial modification is based on a protocol that is called mutant operators. The effectiveness of a test suite in indicating a program change is essential for software testing. Few test cases are more effective than the others, but it is complicated task to identify them. The test can be "good" in the way that it is able of identifying each error that no other test can, but it is complicated to determine because the exposing error is not existing in the code. Otherwise, the test may be "poor" in the way that it is not able to determine any modification, but again this is difficult to detect because authors don't know if any modification is existing in the source code. In each case, there is no technique which can identify the effectiveness of test cases [19]. The author proposed the mutation testing to produce the way of iteratively enhancing test data ability with respect to some program under test (PUT) [1]. The effectiveness outcome of mutation testing is measured when increasing the accuracy of the test data enhanced the tester's confidence in the accuracy of the PUT [27]. Authors have compared and analyzed the performance of six existing open-source MTTs. The results have been validated using 20 JAVA programs along with two real-life applications. The programs were executed on JAVA-based open-source MTTs, and the results were theoretically and experimentally analyzed. The authors highlight practical concepts in an interpreted form that should be effective for both professionals and scientists involved in using mutation tools for mutation testing.

The organization of this chapter includes the following: Sect. 2 describes the related work in the area of mutation testing. Section 3 provides a formulation of

M. Khari (✉)
School of Computer and Systems Sciences,
Jawaharlal Nehru University, Delhi, India

M. Khari et al. (eds.), *Optimization of Automated Software Testing Using Meta-Heuristic Techniques*, EAI/Springer Innovations in Communication and Computing, https://doi.org/10.1007/978-3-031-07297-0_2

research questions. Section 4 presents empirical data collection. Section 5 analyzes the results with the solution of formulated research questions which include descriptive statistics and major issues to be considered while working on open-source mutation testing tools. Section 6 presents the conclusion of the research and projects future scope of this work.

2 Related Work

Mutation testing is the approach which is applicable to determine the thoroughness of the test cases by evaluating the limit to which the test suite can discriminate the program from slight variations of the program. The successful automation of mutation testing leads to decrease in the cost; therefore, it permits mutation testing to be easily capable and effective. The idea of change testing was presented in 1971 in Richard Lipton's research project titled "Fault Diagnosis of Computer Programs." Mutation testing got mainstream in the late 1970s and early 1980s; however, it failed to make the polished product in the industry. Although mutation has appeared very powerful [1–6], it has been unluckily to prove that it is most demanding (high on cost) in the way to produce and execute the mutated programs.

To decrease the final cost, diverse mutation schemes have been introduced in past years considering the respective view. The technique, called "mutant schemata," aims to decrease the production and assortment amount of the generated code [7]. Mostly progress in this field has been explored by DeMillo and Offutt [8], and this progress has been shown in the technique called "constraint-based testing." Various recent advances in mutation testing are symbolic execution [9], concolic execution [10], and search-based optimization techniques [11].

In 2006, Youssef et al. presented a functional verification which contributed to reducing the cost of hardware design flow. This new approach combined mutation-based test techniques and genetic algorithm [24]. In 2009, Schuler et al. presented a flexible approach for benchmarking of mutation testing strategies using tools and concluded not only on effectiveness but also on a comparison of different testing strategies [15].

In 2011, Ferrari et al. discussed three issues associated with the use of mutation investigation and proposed a new model to replace the issue of current faults models. The three tools AjMutator, Proteum/AJ, and MuJava were used for mutation analysis [25]. In 2012, Papadakis described a test case generation method via path selection to kill mutant and analyze mutant. In the same year, artificial insertion of faults under SUT was performed by means of a set of mutation operators along with the empirical study of different combined strategies [18]. In 2012, Kim et al. described the simulation based on JAVA mutation tool, that is, MuJava. Also, in 2012, Dabas et al. considered JAVA as the base language to test the code in mutation testing, and they also used class operators and method operators.

In 2013, a major architecture for mutation analysis and fault seeding technique gives the compiler integrated mutation analyzer for JUnit implementation [13]. Aichernig et al. designed a new model-based test data generation that derives data from the UML state machine [30]. In 2015, Nanavati et al. proposed two new mutation killing criteria, memory fault identification and the control flow deviation, which are useful in decreasing the number of survived mutants [28]. Li et al. implement mutation testing on Ruby programming and develop eight new mutation operators for a Medidata project within an agile development sprint [29, 33, 34].

3 Formulation of Research Questions

Test data generation is very crucial for any software testing. Some of the techniques for generating test data are based on program specifications (known as functional testing) including equivalence class testing, random testing, etc.; on the other hand, others are on the bases of the information about the program code (known as structural testing), including path testing, branch testing, and data flow testing. The test cases were generated on the basis of data flow testing and boundary value analysis of the program. These test cases were applied on 20 JAVA programs and two real-life applications using six JAVA-based MTTs. Finally, the results of these tools (Jester, JavaMut, MuJava, Jumble, Judy, and Javalanche) were compared and validated with the following research questions mentioned in Table 1. Based on the experimental analysis of MTTs, the authors were able to find the best MTT among the selected tools.

4 Empirical Data Collection

This research is based on 20 JAVA programs and two real-life applications. Authors have analyzed and compared the six open-source JAVA-based MTT on the basis of their theory and four quality parameters including memory usage, CPU utilization, mutation score, and the time taken per mutant to kill a fault. The primary commitment of this paper is to look at the flaw's recognition capacity of the test suites. The flaws may be naturally created, or they can be genuine shortcomings. Authors produced mutants from the JAVA programs utilizing a set of standard change mutation operators from the literature on mutation testing.

Table 1 List of research questions

Research questions
RQ1 which MTT, that is, jester, JavaMut, MuJava, jumble, Judy, and Javalanche, is better based on the theoretical comparison?
RQ2 how many operators are supported by each MTT?
RQ3 which MTT is the best in terms of quality, parameters such as memory usage, CPU utilization, mutation score, and the time it took per mutant to kill a fault?
RQ4 what is the statistical evidence to prove the best tool among the six MTT?
RQ5 what are the major issues to be considered while working on six JAVA-based MTT?

Table 2 List of programs

S. No.	Program description	S. No.	Program description
P1	Fibonacci series	P12	Sum of two numbers
P2	Reverse of a number	P13	Prime number next to given number
P3	To find a number is prime or not	P14	Square root of a number
P4	Factorial	P15	Sum of digits
P5	Program to sort an array	P16	Convert temperature from Fahrenheit to Celsius
P6	Print numbers starting from 1 to given number	P17	To find area circumference of circle
P7	Display five numbers randomly	P18	Perimeter of a rectangle
P8	Greater of two numbers	P19	Display digits of a number
P9	Greatest of three numbers	P20	Area of a rectangle
P10	Convert uppercase letters to lowercase and vv	App1	Small text information system
P11	Sum and average of array elements	App 2	Algebraic application

The result suggests that produced mutants were complimentary to real errors, but it is distinct from hand-seeded errors. Also, hand-seeded errors are complicated to identify than real errors. If the output produced from the mutated program is different than the actual output, the mutant is considered killed, or else it is considered alive, that is, live/equivalent. The representation of the six tools throughout the paper is Jester as T1, JavaMut as T2, MuJava as T3, Jumble as T4, Judy as T5, and Javalanche as T6. Table 2 enlists the considered JAVA programs and the real-time applications. The JAVA programs and applications contain approximately 35–1000 lines of source code.

The applications are collected from two different repositories. App 1 is an example of Web application: Small Text Information System (STIS). The source code for app 1 is freely available at [32]. STIS is a nontrivial web application to support users to keep a trail of random textual information. It keeps all the data in the database (MySQL), and it is a combination of 18 JAVA Server Pages and 6 JAVA bean classes. App 2 is available at [33]. This application is named as an algebraic application to give information about the roots of the quadratic expressions. It comprises of two methods: one is a rootFinder with one argument for accepting the discriminant value, and the other is a calDiscriminant with three arguments for accepting constants a, b, and c. It also involves test class TestQuadRoots to kill generated mutants. This test class has two methods: testRootFinder and testCalDiscriminant.

5 Analysis of Results

This section illustrates the output of mutation analysis, addressing each research question mentioned in Sect. 4.

RQ 1. Based on the theoretical analysis, which MTT, that is, Jester, JavaMut, MuJava, Jumble, Javalanche, and Judy, is better?

MTT provides automation to manual mutation testing process which has heavy execution cost. The mutation tools are depending on language because they help mutation operators for the particular programming language. Mothra MTT [12] based on FORTRAN [16], Proteum MTT [13] on C [17] and C++, Jester, and Judy on JAVA [22] and MTTs on Oracle [23] are examples of mutation tool. Our aim is to focus on JAVA language-based mutation testing tools. Authors explore various tools based on mutation testing, but authors found that only a few tools are available as open-source and rest tools like insure++, Certitude, etc. are either commercially licensed or not easily accessible.

T1 [6] is basic JAVA MTT it uses a content based discover and reconstitute approach for producing mutants. T2 [21] was the first executions of the pure JAVA-based mutation framework. It focuses on the JAVA sentence structure and backing mutation operators for object-oriented features and for the conventional mutation operators. T3 [14] supplements T2 tool with extra change mutation operators for JAVA language. Likewise, it backs both area and execution of mutants and additionally gives an instinctive GUI. To decrease the area and simulation amount, one of the T3 forms handles bytecode specifically and utilizes the mutant schemata Scheme [7]. Test cases are provided in a particular configuration, particularly as a JAVA class that holds one technique for every test. Each one test system gives back a string that is utilized to contrast the yields of mutants and the yields of the first class.

There is other JAVA MTT that supports JUnit test cases: Testooj, Jumble, Judy, and Javalanche. The Testooj is the test data production tool for JAVA-based language. It coordinates a few existing testing devices including T3, and it permits T3 to utilize JUnit test cases. T4 [3] specifically handles bytecodes to accelerate mutation testing. Notwithstanding, with the exception of producing mutants, it utilizes the same approach as T5. Likewise, it doesn't help class mutation operators. T5 [4] utilizes aspect-oriented mechanisms to stay away from various assemblages of mutants. This executes both traditional and class mutation operators. T6 [15] is a moderately novel JAVA-based mutation tool. It receives a few expense diminishment approaches. For instance, it controls bytecode specifically and utilizes a mutant schemata procedure [17]. Also, it executes just those test cases that are well known to cover the mutated statement by using coverage data.

Table 3 provides a comparison of the MTTs in terms of year of introduction, generation of tools, the number of traditional and class operator supported, GUL/CUI category, and their JUnit support availability and key characteristics.

RQ 2. How many operators are supported by each MTT?

Each MTT has different sets of mutant operators. In literature five types of mutation operators are available, namely, conventional mutation operators, class mutation operators, state-based mutation operators [22], method level mutation operators, and mutation operators for JAVA-based specific features. T3 is considered to be the best as it supports a wide range of operators like five traditional operators, 24 class level operators, and method level operators [26]. A detailed comparison of all the six tools is mentioned in Table 3. Table 4 includes a lists of the mutation operators supported by selected MTTs and their abbreviations.

Table 3 Comparison of six mutation testing tools based on JAVA language

Tool	Year of introduction	Generation	No. of traditional operators	No. of class operators	Mutant format	GUI/CLI	JUnit support	Characteristics
T1 [6]	2001	Source	2	Nil	Separate source files	GUI	JUnit4	Produced web pages displaying the outputs of the tests by using the built-in script
T2 [20]	2002	Source	6	20	In separate class file	CLI	JUnit3	First fully fledged JAVA mutation system, base of MuJava
T3 [14, 20, 21]	2005	Source	5	24	Separate class file	GUI	JUnit3	Weak mutation, mutant schemata, reflection technique
T4 [3]	2007	BCEL	7	Nil	In memory	GUI	JUnit4	Class level mutation testing tool
T5 [4, 31]	2010	Source	5	7	Grouped in source files	GUI	JUnit4	Support for the latest version of JAVA, traditional and class mutation operators
T6 [15]	2009	ASM	5	Nil	Separate class file	GUI	JUnit4	Invariant and impact analysis
Needs	Recent	Source	>5	>20	In separate class file	GUI	Yes	

Table 4 List of abbreviations and their supported tools

Operator	Description	Tools	Operator	Description	Tools
ABS	Absolute value insertion	T3, T5	PNC	New method calls with child class type	T3
LCR	Logical connector replacement	T3, T5	PMD	Member variable declaration with parent class type	T3
ROR	Relational operator replacement	T1, T2, T3, T4, T5	PPD	Parameter variable declaration with child class type	T3
UOI	Unary operator insertion	T3, T5	PCI	Type cast operator insertion	T3
AOR	Arithmetic operator replacement	T3, T4, T5	PCD	Typecast operator deletion	T3
ERP	Event-keyword replacement	T3	PCC	Typecastoperator to change	T3
EDL	Event-keyword deletion	T3	PRV	Reference assignment with another comparable variable	T3
MNR	Method name replacement	T3	OMR	Overloading method contents replace	T3
MND	Method name deletion	T3	OMD	Overloading method deletion	T3
DNR	Data name replacement	T3	OAC	Arguments of overloading method call change	T3
DVR	Data value replacement	T3	JDC	JAVA-supported default constructor create	T3
DVN	Data value for data names replacement	T3	AMC	Access modifier change	T3
ROR	Relational operator replacement	T3	IHI	Hiding variable insertion	T3
OAR	Arithmetic operator replacement	T3	IHD	Hiding variable deletion	T3
AOI	Arithmetic operator insertion	T3	IOD	Overriding method deletion	T3
AOD	Arithmetic operator deletion	T3	IOP	Overriding method calling position change	T3
COR	Conditional operator replacement	T2,T3, T4,T5	IOR	Overriding method renames	T3
COI	Conditional operator insertion	T1,T2,T3	ISD	Super keyword deletion	T3
COD	Conditional operator deletion	T1,T2,T3	ISI	Super keyword insertion	T3
SOR	Shift operator replacement	T2,T3, T4,T5	IPC	Explicit call of a parent's constructor deletion	T3
LOR	Logical operator replacement	T2,T3, T4,T5	JSC	Static modifier change	T3

(continued)

Table 4 (continued)

Operator	Description	Tools	Operator	Description	Tools
LOI	Logical operator insertion	T2,T3	JID	Member variable initialization deletion	T3
LOD	Logical operator deletion	T2,T3	JTD	This keyword deletion	T3,T5
ASR	Assignment operator replacement	T2,T3, T5	UOD	Unary operator deletion	T5
EOA	Reference assignment and content assignment replacement	T5	EOC	Reference comparison and content comparison replacement	T5
JTI	This keyword insertion	T5	EAM	Accessor method change	T5
EMM	Modifier method change	T5	RNC	Replace numerical constant	T6
RAO	Replace arithmetic operator	T6	OMC	Omit method calls	T6
AORB	Arithmetic operator replacement binary	T2	AORU	Arithmetic operator replacement unary	T2
AORS	Arithmetic operator replacement shortcut	T2	AOIU	Arithmetic operator insertion unary	T2
AOIS	Arithmetic operator insertion shortcut	T2	AODU	Arithmetic operator deletion unary	T2
AODS	Arithmetic operator deletion shortcut	T2	AVI	Absolute value insertion	T4

RQ 3. Which MTT is the best in terms of quality parameters, such as memory usage, CPU utilization, mutation score, and the time took per mutant to kill a fault?

The quality of MTTs is measured in terms of memory usage, CPU utilization, mutation score, and time. All these parameters are discussed in this RQs.

Memory Usage: The memory usage refers to the memory used by a program during execution. The authors have focused on six MTTs and calculated memory consumption for particular and each PUT. The output of memory consumption is presented in Table 5, and the graph is presented in Fig. 1. After analysis, performance of T3 is the best in program P1. Mutation tool T3 consumes 12.09 MB memory, and other tools are consuming T2-16.11, T3-28.74, T4-23.2, T5-24.05, and T6-27.29. The performance of all subject programs with respect to the memory consumption with tools is T1-22.7, T2-26.2, T3-11.6, T4-21.5, T5-23.8, and T6-25. The T3 tool is performed well best as compared to other tools.

The estimated memory usage and the corresponding graphical representation for the six MTTs are given in Table 5 and Fig. 1, respectively. Clearly, T3 uses the least memory in comparison to other MTTs with the average memory usage of 11.6 (averages of T1, T2, T4, T5, and T6 are 22.7, 26.2, 21.5, 23.8, and 25, respectively).

CPU Utilization: CPU usage alludes to a workstation's use of transforming assets, or the measure of work taken care of by a CPU. Real CPU use differs upon the sum and kind of oversaw processing errands. Certain errands oblige substantial CPU time, while others require less as a result of non-CPU asset prerequisites. The

Table 5 Memory usage of six mutation testing tools

	T1	T2	T3	T4	T5	T6		T1	T2	T3	T4	T5	T6
P1	16.11	28.74	11.09	23.21	24.05	27.29	**P12**	21.45	23.37	13.35	19.36	23.24	27.26
P2	18.90	28.24	10.12	20.98	21.87	25.19	**P13**	26.27	24.12	12.14	19.61	24.21	27.66
P3	24.22	27.77	13.62	19.96	24.06	26.37	**P14**	24.55	24.39	12.34	20.48	25.62	27.00
P4	24.07	27.02	12.36	25.34	23.56	20.96	**P15**	23.93	25.03	11.15	19.20	24.91	27.69
P5	23.04	25.81	10.04	25.27	22.85	24.21	**P16**	17.22	25.52	10.03	18.43	23.82	20.98
P6	24.34	27.69	12.07	20.85	22.91	21.14	**P17**	18.77	27.27	11.85	27.21	26.30	24.31
P7	21.61	25.75	10.4	19.50	24.05	21.63	**P18**	25.31	25.81	9.95	21.59	24.54	24.42
P8	23.15	26.69	12.9	20.44	22.59	23.39	**P19**	25.03	27.05	12.8	23.10	23.85	25.27
P9	21.91	28.25	12.61	24.17	21.68	27.20	**P20**	26.34	25.82	12.41	17.91	22.13	25.39
P10	21.42	27.88	11.43	17.11	23.76	26.35	**App. 1**	26.79	27.77	12.04	26.71	26.71	27.71
P11	22.25	23.81	12.04	19.59	24.06	23.76	**App. 2**	24.61	22.89	10.24	24.07	24.84	26.56

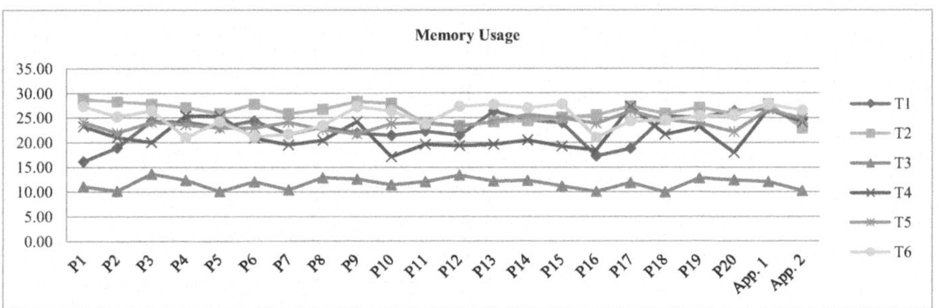

Fig. 1 Graphical representation of results of memory usage

Table 6 CPU utilization of six mutation testing tools

	T-1	T-2	T-3	T-4	T-5	T-6		T-1	T-2	T-3	T-4	T-5	T-6
P1	16.41	24.09	31.01	24.32	14.02	24.31	**P12**	17.61	26.34	30.08	23.21	12.19	21.82
P2	12.63	23.04	30.61	23.12	13.91	23.62	**P13**	18.98	25.32	32.02	24.72	18.32	24.59
P3	13.22	23.91	32.66	24.26	12.61	25.81	**P14**	13.96	26.87	31.91	24.17	18.61	21.66
P4	11.74	18.82	30.13	22.10	14.33	26.98	**P15**	14.81	25.62	31.83	22.61	17.64	24.88
P5	13.91	21.63	30.16	22.76	14.32	21.68	**P16**	19.28	23.77	32.01	23.48	16.79	21.17
P6	14.14	22.19	32.42	23.22	13.61	22.94	**P17**	17.49	24.91	32.64	23.19	16.33	21.83
P7	13.72	22.64	32.14	22.01	13.19	20.15	**P18**	16.71	24.92	31.27	24.02	15.91	21.52
P8	14.68	22.72	33.06	23.57	11.31	21.43	**P19**	19.21	26.74	33.45	24.91	15.32	24.99
P9	14.72	24.81	31.15	24.12	12.98	23.64	**P20**	19.47	26.72	30.61	23.60	13.14	21.17
P10	11.24	23.99	30.41	23.14	13.17	23.49	**App. 1**	19.43	27.62	35.65	22.32	20.19	26.28
P11	17.32	25.62	33.74	22.67	14.32	24.79	**App. 2**	16.12	30.08	33.81	20.09	19.32	22.59

CPU performs all the evaluations that are required to procedure transactions. The huge transaction-related estimations that it works inside a specific period, the increases the throughput will be for that particular period. The results of CPU utilization are represented in Table 6, and the graphical representation is depicted in Fig. 2. The average performance of all subject programs with respect to the CPU

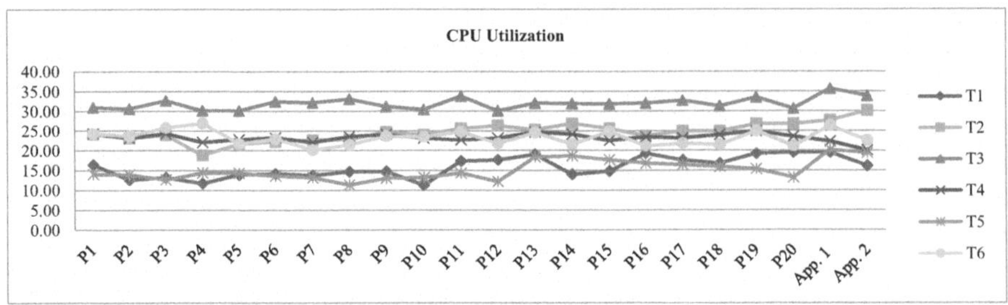

Fig. 2 Graphical representation of results of CPU utilization

Table 7 Mutation score of six mutation testing tools

	T1	T2	T3	T4	T5	T6		T1	T2	T3	T4	T5	T6
P1	68%	18%	68%	43%	37%	32%	**P12**	35%	34%	24%	19%	15%	35%
P2	62%	24%	57%	29%	16%	37%	**P13**	39%	32%	31%	39%	38%	39%
P3	63%	14%	52%	40%	17%	33%	**P14**	40%	32%	58%	26%	20%	40%
P4	65%	21%	60%	13%	28%	19%	**P15**	56%	40%	30%	37%	32%	56%
P5	63%	20%	63%	37%	14%	15%	**P16**	32%	23%	78%	82%	55%	32%
P6	61%	35%	61%	21%	12%	17%	**P17**	44%	28%	82%	88%	75%	44%
P7	52%	12%	36%	25%	9%	14%	**P18**	41%	38%	56%	60%	45%	41%
P8	59%	22%	32%	35%	14%	36%	**P19**	40%	24%	21%	12%	11%	40%
P9	64%	13%	64%	25%	5%	16%	**P20**	43%	23%	19%	18%	15%	43%
P10	68%	20%	68%	21%	14%	5%	**App. 1**	87%	70%	81%	58%	39%	87%
P11	56%	47%	50%	36%	47%	43%	**App. 2**	54%	54%	72%	47%	49%	54%

utilization through tools T1, T2, T3, T4, T5, and T6 is 15.7%, 24.6%, 31.9%, 23.2%, 15%, and 23.2%, respectively. On an average, the performance of T3 is best in comparison to the other considered tools.

Mutation Score: The live mutants are those mutants which are not processed even once. The live mutant depicts one of the following two situations. First, the modified project is directly proportionate to the first practically. Second is unhiding the experiments that don't present or the test case is incompetent for unrevealing and assassinating it. So, authors focused on similar mutant as the alive mutant, similar mutants generate the same result as the PUT always, nevertheless of the test supplied.

Mutants are killed by iteratively producing and executing test cases, in order to determine the killed mutants. The determination of a program to be live and killed is done by differentiating the result of the original and mutated programs. Authors have executed each program on each and every tool, and their respective results of mutation score are shown in Table 7, and the graphical representation is depicted in Fig. 3. Total numbers of mutants are always equal to the combination of live mutants and killed mutants. Authors calculate the mutation score (M_S) where killed mutant (M_K), total mutant (M_T), and live/equivalent mutants ($M_{L/E}$) are based on Eq. 1:

$$M_S = M_K / M_T \left(M_K + M_{L/E}\right) \tag{1}$$

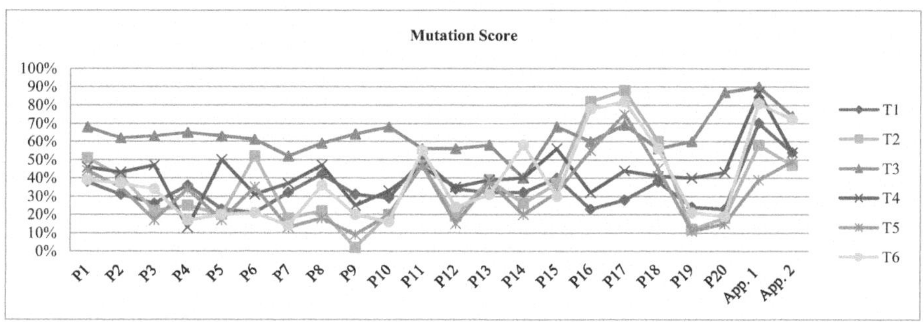

Fig. 3 Graphical representation of results of mutation score

Table 8 Time was taken to kill one mutant by six mutation testing tools

	T1	T2	T3	T4	T5	T6		T1	T2	T3	T4	T5	T6
P1	12.12	7.07	1.85	5.38	13.12	14.22	**P12**	0.22	0.25	0.03	0.05	0.59	0.39
P2	1.84	1.24	0.47	0.99	2	2.22	**P13**	11.89	8	2.5	6.58	7.98	10.8
P3	2.07	1.28	0.14	0.41	2.15	0.98	**P14**	3.3	4.25	1.26	1.96	5.49	2.56
P4	6	16.5	2.62	16.08	6.39	23.78	**P15**	0.22	0.75	0.04	0.17	1.39	0.88
P5	17.42	20.38	2.32	5.91	14.16	11	**P16**	0.92	0.36	0.04	0.29	1	0.44
P6	4.05	1.3	0.67	2.68	3.14	4.67	**P17**	0.45	0.12	0.03	0.1	0.18	0.21
P7	0.97	1.74	0.22	0.56	3.45	3.1	**P18**	1.94	1.22	0.47	1.66	0.8	3.52
P8	0.17	0.25	0.03	0.08	0.6	0.19	**P19**	0.76	0.59	0.06	0.27	1.68	0.93
P9	1.87	32.7	0.62	2.4	17.95	2.77	**P20**	0.06	0.23	0.01	0.02	0.12	0.14
P10	0.37	0.19	0.02	0.25	0.6	0.82	**App. 1**	1.68	2.53	0.82	1.09	4.08	2.07
P11	0.69	0.96	0.3	0.99	0.89	0.84	**App. 2**	1.38	0.51	0.23	1.06	1.3	0.7

The average performance of all subject programs with respect to the MS through tools T1, T2, T3, T4, T5, and T6 is 34%, 37%, 64%, 42%, 31%, and 39%, respectively. The average performance of T3 in terms of MS is best compared to other considered tools.

Time is taken to kill one mutant:

Now, evaluate the total time required to execute a particular program by a subjected tool and then calculate the kill rate of a mutant on a particular tool. With the support of time in MS and the information about the total killed mutant, the authors calculate the time it takes to kill one mutant. TM_K stands for time per mutant, where Time (T) is in Eq. 2:

$$TM_K = T / M_K \tag{2}$$

The results of the time taken to kill one mutant are represented in Table 8, and the graphical representation is depicted in Fig. 4. The average performance of all subject programs with respect to the time taken to kill one mutant through tools T-1, T-2, T-3, T-4, T-5, and T-6 is 6.6, 6.5, 1.8, 5.8, 9.3, and 9.6, respectively. Tool T3 takes less time in terms of TM_K compared to other considered tools.

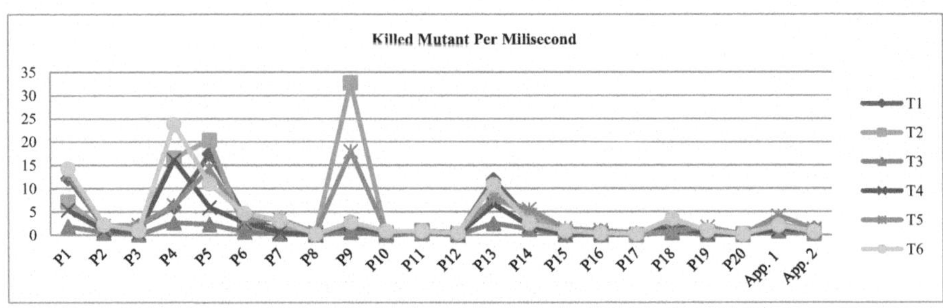

Fig. 4 Graphical representation of results of time taken to kill one mutant

Table 9 Average performance of each mutation testing tools versus parameters

Parameters/tools	T1	T2	T3	T4	T5	T6
Memory usage	22.79	26.21	11.68	21.55	23.89	25.08
CPU utilization	15.76	24.65	31.94	23.26	15.07	23.24
Mutation score	34	37	64	42	31	39
Kill rate	3.2	4.6	0.67	2.23	4.05	3.97

RQ 4. What is the statistical evidence to prove the best tool among the six MTT?

Statistical evidence to prove the best tool among the six MTTs is the average performance of each tool in terms of memory usage, CPU utilization, mutation score, and kill rate, which is shown in Table 9.

First, calculate the standard deviation (σ) of each parameter using the "n-1" method. Authors retrieve σ 9.82 for memory usage, for CPU utilization – 6.24, for mutation score – 11.8, and for kill rate σ 1.45. σ this symbol is considered that its arguments is the sample of the population.

$$\sigma = \sqrt{\frac{\left(x-\overline{x}\right)^2}{\left(n-1\right)}} \tag{3}$$

Here x is the sample mean, and n is the sample size. Example mean, x, is at the focal point of this extent, and the reach is confidence interval.

For instance, if x is the mean of CPU utilization for items, then confidence interval is an extent of the population mean on the basis of the sample mean.

For each population means μ, the likelihood of getting a specimen mean further from μ than x is more stupendous than alpha (the level of significance). For each population mean, μ, in this extent, the likelihood of acquiring an example mean additionally from μ than x is short of what alpha. As it were, expect that authors utilize x, which is the standard deviation and size to develop a two-tailed test criticalness level alpha of the speculation that the population mean is μ. At that point, the authors won't dismiss that assumption if μ is in the confidence interval and will dismiss that theory if μ is not in the confidence period. The confidence period does not permit us to surmise the likelihood that the next bundle will take over the conveyance time that is in the confidence period. Here, alpha = 0.05, it is needed to evaluate the area under the standard normal curve that equals or 95%. This value is ±1.96. The confidence interval is calculated by using the given formula in Eq. 4 (Table 10):

$$\bar{x} \pm 1.96\left(\frac{\sigma}{\sqrt{n}}\right) \tag{4}$$

For example, for tool T1, the confidence interval for the memory usage is lying between 21.57 and 24.01. It means the memory usage of the tool is between 21.57 and 24.01 at 95% of the time. If authors take any 100 programs, then 95 programs will be having the memory usage lying in the given interval.

RQ 5. What are the major issues to be considered while working on six JAVA-based MTT?

While working with all the six JAVA-based MTT, few issues have occurred, and they must be considered at the time of practical usage by software practitioners and researchers. These are mentioned below:

- Major Issues in T1: T1 scheme is to produce, compile, and execute unit tests across the mutant. The procedure is repeated for each mutant of source program, and, therefore, it is not efficient. Hence, a long time is taken by T1 to run, and the output requires little man-made effort to change. It does not support the OO mutation operators.
- Major Issues in T2: It supports only 20 OO operators for mutation. It supports command line interpreter.
- Major Issues in T3: Representation of output is not user-friendly.
- Major Issues in T4: OO mutation operators are not supported by the Jumble tool. Fixed replacements for the other mutation operators are provided by this tool.
- Major Issues in T5: Every generated mutant is not compiled. So, there is a tendency to miss out changes in complex mutants that might have been unique to them. JVM limits the size of the class. Mass compilation of mutants was not effective. A compilation error generated by a single mutant breaks the processing of all simultaneously compiled mutants.
- Major Issues in T6: Scalability problem occurred while executing 500 LOC, and hence it gets hanged up. Running T6 on a library shows that it has dependencies (e.g., XStream) that can become a tricky issue. Build files are missing; hence, they have to be downloaded from the Linux software center. If one has to run the source code for a mutation, then that code has to be copied from the Javalanche folder. It does not support the OO mutation operators.

Table 10 95% confidence interval for mean

Parameter/tools	T1	T2	T3	T4	T5	T6
Memory usage	(24.01, 21.57)	(26.93,25.49)	(12.16,11.20)	(22.76,20.34)	(24.44,23.35)	(26.04,24.12)
CPU utilization	(16.85,14.68)	(25.65,23.66)	(32.54,31.35)	(23.70,22.81)	(16.11,14.03)	(24.03,22.45)
Mutation score	(0.39,0.30)	(0.46,0.27)	(0.68,0.59)	(0.48,0.36)	(0.38,0.24)	(0.50,0.29)
Kill rate	(5.14,1.26)	(8.10,1.21)	(1.03,0.31)	(3.76,0.70)	(6.14,1.96)	(6.43,1.50)

6 Conclusions and Future Work

Mutation testing is the artificial seeding of faults into the PUT. The authors have analyzed six open-source JAVA-based MTT (Jester, JavaMut, MuJava, Jumble, Judy, and Javalanche). The tools have been compared theoretically and empirically, and the results were formulated. Benchmark comparison among the MTT is presented for issues in terms of mutants, mutation operator, mutation score, and quality of output. On the basis of performance analysis, each tool is discussed along with the protocols for identifying the best tool among the six JAVA-based MTT in terms of killing the mutants. The results show that the MuJava, that is, T3 MTT, performs best when compared to T1, T2, T4, T5, and T6. The research questions and their solutions provide guidelines for selecting MTT for practitioners and researchers.

Future scope of our research is to automate the production of mutants and test cases that will provide the solution to the mutation issues to some extent and introduce some other OO mutation operators, and it can be helpful to produce mutants that are killed by the current test data set. Additional research will be done by using case studies to differentiate mutation operators, and it provides real hints on how to select appropriate operators.

References

1. DeMillo, R. A., Lipton, R. J., & Sayward, F. G. (1978). Hints on test data selection: Help for the practicing programmer. *Computer, 11*(4), 34–41.
2. Jeevarathinam, R., & Thanamani, A. S. (2011). *A survey on mutation testing methods, fault classifications and automatic test cases generation.*
3. Irvine, S. A., Pavlinic, T., Trigg, L., Cleary, J. G., Inglis, S., & Utting, M. (2007, September). Jumble java byte code to measure the effectiveness of unit tests. In *Testing: Academic and industrial conference practice and research techniques-MUTATION (TAICPART-MUTATION 2007)* (pp. 169–175). IEEE.
4. Madeyski, L., & Radyk, N. (2010). Judy – A mutation testing tool for Java. *IET Software, 4*(1), 32–42.
5. Offutt, A. J., Pan, J., Tewary, K., & Zhang, T. (1996). An experimental evaluation of data flow and mutation testing. *Software: Practice and Experience, 26*(2), 165–176.
6. Moore, I. (2001). Jester-a JUnit test tester. *Proceedings of 2nd XP*, 84–87.
7. Untch, R. H., Offutt, A. J., & Harrold, M. J. (1993, July). Mutation analysis using mutant schemata. In *Proceedings of the 1993 ACM SIGSOFT international symposium on Software testing and analysis* (pp. 139–148).
8. DeMillo, R. A., & Offutt, A. J. (1991). Constraint-based automatic test data generation. *IEEE Transactions on Software Engineering, 17*(9), 900–910.
9. King, J. C. (1976). Symbolic execution and program testing. *Communications of the ACM, 19*(7), 385–394.
10. Sen, K., Marinov, D., & Agha, G. (2005). CUTE: A concolic unit testing engine for C. *ACM SIGSOFT Software Engineering Notes, 30*(5), 263–272.
11. Harman, M., & McMinn, P. (2007, July). A theoretical & empirical analysis of evolutionary testing and hill climbing for structural test data generation. In *Proceedings of the 2007 international symposium on software testing and analysis* (pp. 73–83).

12. Zapf, C. N. (1993). *MedusaMothra-A distributed interpreter for the Mothra mutation testing system* (Master's thesis, Clemson University).
13. Maldonado, J. C., Delamaro, M. E., Fabbri, S. C., da Silva Simão, A., Sugeta, T., Vincenzi, A. M. R., & Masiero, P. C. (2001). Proteum: A family of tools to support specification and program testing based on mutation. In *Mutation testing for the new century* (pp. 113–116). Springer.
14. Ma, Y. S., Offutt, J., & Kwon, Y. R. (2006, May). MuJava: A mutation system for Java. In *Proceedings of the 28th international conference on Software engineering* (pp. 827–830).
15. Schuler, D., & Zeller, A. (2009, August). Javalanche: Efficient mutation testing for Java. In *Proceedings of the 7th joint meeting of the European software engineering conference and the ACM SIGSOFT symposium on the foundations of software engineering* (pp. 297–298).
16. Offutt, A. J., & King, K. N. (1987, July). A Fortran 77 interpreter for mutation analysis. In *Papers of the symposium on interpreters and interpretive techniques* (pp. 177–188).
17. Agrawal, H., DeMillo, R., Hathaway, R., Hsu, W., Hsu, W., Krauser, E. W., ... Spafford, E. (1989). *Design of mutant operators for the C programming language* (Technical report SERC-TR-41-P). Software Engineering Research Center, Purdue University.
18. Papadakis, M., & Malevris, N. (2012). Mutation based test case generation via a path selection strategy. *Information and Software Technology, 54*(9), 915–932.
19. Delamaro, M. E., Maldonado, J. C., Pasquini, A., & Mathur, A. P. (2001). Interface mutation test adequacy criterion: An empirical evaluation. *Empirical Software Engineering, 6*(2), 111–142.
20. Chevalley, P., & Thevenod-Fosse, P. (2003). A mutation analysis tool for Java programs. *International Journal on Software Tools for Technology Transfer, 5*(1), 90–103.
21. Guderlei, R., Just, R., Schneckenburger, C., & Schweiggert, F. (2008, April). Benchmarking testing strategies with tools from mutation analysis. In *2008 IEEE international conference on software testing verification and validation workshop* (pp. 360–364). IEEE.
22. Alexander, R. T., Bieman, J. M., Ghosh, S., & Ji, B. (2002, November). Mutation of Java objects. In *13th international symposium on software reliability engineering, 2002. Proceedings* (pp. 341–351). IEEE.
23. Tuya, J., Suárez-Cabal, M. J., & De La Riva, C. (2007). Mutating database queries. *Information and Software Technology, 49*(4), 398–417.
24. Serrestou, Y., Beroulle, V., & Robach, C. (2006, April). How to improve a set of design validation data by using mutation-based test. In *2006 IEEE design and diagnostics of electronic circuits and systems* (pp. 75–76). IEEE.
25. Ferrari, F. C., Nakagawa, E. Y., Maldonado, J. C., & Rashid, A. (2011, March). Proteum/AJ: A mutation system for AspectJ programs. In *Proceedings of the tenth international conference on Aspect-oriented software development companion* (pp. 73–74).
26. Ma, Y. S., Harrold, M. J., & Kwon, Y. R. (2006, May). Evaluation of mutation testing for object-oriented programs. In *Proceedings of the 28th international conference on software engineering* (pp. 869–872).
27. Jia, Y., & Harman, M. (2010). An analysis and survey of the development of mutation testing. *IEEE Transactions on Software Engineering, 37*(5), 649–678.
28. Nanavati, J., Wu, F., Harman, M., Jia, Y., & Krinke, J. (2015). Mutation testing of memory-related operators. In *2015 IEEE eighth International Conference on Software Testing, Verification and Validation Workshops (ICSTW)* (pp. 1–10). IEEE.
29. Li, N., West, M., Escalona, A., & Durelli, V. H. (2015, April). Mutation testing in practice using ruby. In *2015 IEEE eighth International Conference on Software Testing, Verification and Validation Workshops (ICSTW)* (pp. 1–6). IEEE.
30. Aichernig, B. K., Brandl, H., Jöbstl, E., Krenn, W., Schlick, R., & Tiran, S. (2015). Killing strategies for model-based mutation testing. *Software Testing, Verification and Reliability, 25*(8), 716–748.

31. Just, R. (2014, July). The major mutation framework: Efficient and scalable mutation analysis for Java. In *Proceedings of the 2014 international symposium on software testing and analysis* (pp. 433–436).
32. http://www.ise.gmo.edu:8080/ofut/jsp/stis
33. https://adrive.com/
34. Khari, M., Dalal, R., & Rohilla, P. (2020). Extended paradigms for botnets with WoT applications: A review. *Smart Innovation of Web of Things, 105.*

State Traversal: Listen to Transitions for Coverage Analysis of Test Cases to Drive the Test

Sonali Pradhan, Mitrabinda Ray, Sukant Bisoyi, and Deepti Bala Mishra

1 Introduction

Testing assesses the quality of the product that is done during the development phase. Software testing is a verification and validation process. Testing at the early phase (design phase) observes early fault detection at which decreases cost in the software development life cycle. The design phase needs to be enhanced to recover the detected faults. The faults which are detected in the design phase need to be recovered before the program is written [1, 2]. To find the effectiveness of test cases and to minimize the development effort, model-based testing (MBT) is a very constructive one [3, 4]. Test cases are generated at the early phase of development using this testing approach [5, 6]. To generate test cases in the design phase, the tester uses unified modeling language (UML) diagrams [7–9].

Initially, the testers went for code coverage testing. In the white box testing, testing of implementation code is done as code-based testing [10–13]. However, in code-based testing, it is difficult to achieve state coverage [14, 15]. In state-based testing (SBT), test scenarios are generated from the state chart diagrams. The coverage analysis is also performed from the source code using various tools. Different

S. Pradhan (✉) · S. Bisoyi
Department of Computer Science and Engineering, C. V. Raman Global University, Bhubaneswar, India

M. Ray
Department of Computer Science and Engineering, S'O'A, Deemed to be University, Bhubaneswar, India
e-mail: mitrabindaray@soa.ac.in

D. B. Mishra
Department of MCA, GITA Autonomous College, Bhubaneswar, India

M. Khari et al. (eds.), *Optimization of Automated Software Testing Using Meta-Heuristic Techniques*, EAI/Springer Innovations in Communication and Computing, https://doi.org/10.1007/978-3-031-07297-0_3

state-based faults are achieved in SBT. The disadvantages we find from the code-based testing over model-based testing are as follows:

- In code-based testing, the dynamic behavior of the system can't be achieved; here state and transition coverage can't be reached.
- In order to perform code-based testing on an application, a tester needs to know the internal structure of the program. However, the code is not always available to the tester.
- To test all possible values of a feature is not possible when the source code is large enough.
- Code-based testing is also not telling how much is the coverage in the code and how well the logic is covered.
- The complexity of an object-oriented system can be remarked by the object interactions. Complex behaviors are observed when related objects pass messages with each other within a scenario, and this type of behavior is achieved in a sequence diagram. In the case of sequence diagrams, interactions between objects are in a timely sequence manner.
- UML sequence diagram and collaboration diagram provide a way to model the behavior of an object-oriented system. The advantages of model-based over code-based are easier extraction of concurrent control flow and the nature of polymorphism.

The main advantage in model-based testing is that the system behavior can be achieved by following the pictorial or tabular representation of the model. It reduces the testing interval without reducing quality as at the early phase of design; it reduces the faults, which saves time and resources. To understand the advantage of model-based testing over code-based testing, we have considered a small example of a source code which is given below:

```
Source Code
Input: a, b
Output: Positive result/Negative result
Step 1: Print (int a, int b)
Step 2: {
Step 3: int result = a + b;
Step 4: If (result > 0)
Step 5: Print ("Positive result", result)
Step 6: Else
Step 7: Print ("Negative result", result)
Step 8 :}
```

Let us say, we skip to check and print the negative part in the source code. Then the following source code will be given below:

```
Source Code:
Input: a, b
Output: Positive result/negative result
Step 1: Print (int a, int b) {
Step 2: int result = a + b;
Step 3: If (result > 0)
Step 4: Print ("Positive result", result)}
```

To test the code coverage mentioned above, we have considered Scenario 1.

```
Scenario 1:
If a=3, b=9

Step 1: Prints (int a, int b) {
Step 2: int result = a + b;
Step 3: If (result > 0)
Step 4: Print ("Positive result", result)
Step 5 :}
```

The statements mentioned above are those which are executed as per the scenario. According to this statement execution, code-based testing is also not telling how much and how well the tester has to cover the logic. If the function is not included in the specification, the structure base testing cannot find that issue. Scenarios are a means of analyzing applications and understanding the dynamic behavior. Another way to model scenarios is by using sequence diagram and collaboration diagram.

State-based testing is a type of black box testing, where early testing activities are done before implementation of the model [7, 15, 16]. In state-based testing, unit testing is done, and the unit's behavior is achieved. It compares its actual states to the expected states. This approach is accepted with object-oriented systems and for complex behavior, such as critical real-time application as a computer system, making a web application to a state chart for designing test cases, effective test strategies for enterprise critical applications, etc. State-based testing also accomplishes state-transition testing where some aspect of the system can be described known as a finite state machine (FSM) [4, 17]. The finite state machine is very much important in software testing. FSM validates an abstract model of the system, tests case generation at the development stage, assigns a pass/fail verdict, analyzes the execution result to enhance the model, reduces cost by updating test cases, and prevents state-based faults.

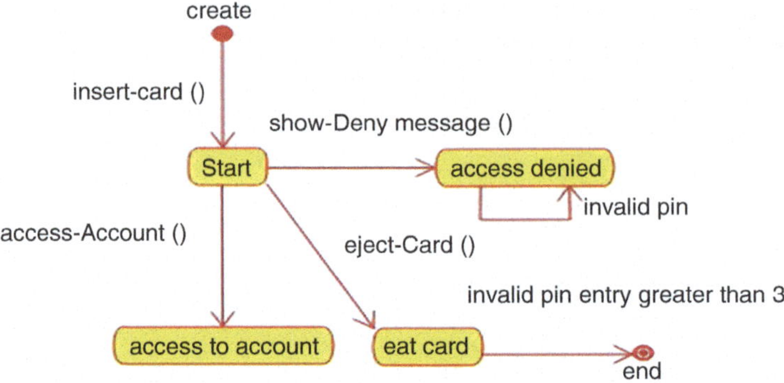

Fig. 1 State diagram for PIN entry transaction

A finite state system is often shown as a state diagram. State-based testing is also a type of model-based testing [18, 19] where the state changes are observed in system under test (SUT). The tester observes all coverage like transition, state, action or event, etc. A state chart diagram is shown in Fig. 1 as personal identity number (PIN) transaction for a bank account. The states are shown in oval shapes, and the transitions are lines with arrows. The events are given the text near the transitions which exhibits triggering a transition when an event occurred. The state diagram shows four states of the transaction such as *start*, *access denied*, *access to account,* and *eat card*, and four possible events are card inserted, entering valid pin, entering invalid pin, and invalid pin entry greater than 3. The state chart diagram shows four events such as insert-Card (), show-Deny message (), access-Account (), and eject-Card ().

To derive test cases, the typical scenarios are as follows:

T1 (test case 1) = insert-Card () ---- access-Account () [start – access to account]
T2 = insert-Card () ---- show Deny message () [start – access denied]
T3 = insert-Card () ----show Deny message () ---- show Deny message () ---- show Deny message () ---- eject Card () [start – eat card]

In state transition technique, a single state can be used to the requirement; otherwise, a series of different states can be used in the model. In case of failure, the suspect area is limited. Identical inputs are not always accepted, and if accepted they may produce different outputs, and this can check the state-based behavior of the system.

For a nonfinite system, where the execution is not in sequential order, it is challenging to get the exact state change in the model behavior. Coverage analysis is always done from the source code. Another greatest disadvantage is that for small systems it is quite simple to define all possible states, but it is not as easy for a larger system as there is an exponential progression in the number of states. Unit (class) testing is a level of software testing where an individual unit of software is tested. In state-based testing, unit/class testing is done. The state-based unit testing makes sure the class behaves as per the design. The purpose of unit testing in state-based is to validate each unit/component of the software with consideration of the design. Chow [20] proposed W-method to generate test sequences from a spanning tree to the behavior of the SUT. W-method is again changed by Binder [15]. There round trip path (RTP) is used as a coverage criterion of UML diagram. Binder proposed a new method for state-based testing. To cover all paths in the diagram, the transition tree is generated, and that is traversed to cover all paths/transitions present in that tree. The repeated state is the stopping state which is already covered when traversed. Then some new coverage criteria are proposed by Turner [21] on state-based testing. Most studied coverage criteria are full predicate (FP), round trip path (RTP), and all transitions (AT), all transition pairs (ATP), ATP paths of length 2 (LN2), ATP paths of length 3 (LN3), and ATP paths of length 4 (LN4) [5, 20, 22]. Some critical aspects of MBT approaches are presented in the survey paper [17]. This chapter elaborates the concepts and frame for accessing model-based testing tools. An industrial testing aspects influence the cost-effectiveness using SBT [4, 17]. In a work [18, 20], they assessed test selection criteria using extended finite state machines (EFSMs). In a broad overview, the tester goes for synchronous and asynchronous testing using behavioral state chart diagram.

The principal aim of the tester is to generate a set of useful test cases for different applications in software development. Most of the literature survey shows the improvement of the algorithm to generate test cases using various types of UML diagrams. The aim is always to generate effective test cases. To generate a minimal set of effective test cases is the most significant challenge. This motivates us to work on state-based testing which deals in synchronous environment. For a large and complex system, manually testing is a very tedious process and time-consuming. Hence, manual testing is not acceptable; the tester always tries for automated test case generation, which is very much helpful in software development. Some little change in the software leads to a significant change in the program. Various works are already done in model-based testing, but the primary challenge is to different state-based faults. The source code may not be available to the tester in all cases; it is needed to generate test cases based on design specifications. When the system under test is running under the asynchronous environment, the sequence of input-output is difficult to achieve [23]. Hence, knowing the proper sequence of input-output is a considerable challenge in asynchronous testing. Some web-based applications are the best way to accomplish asynchronous state-based testing. Hence, our work is motivated by the requirement to address the issues mentioned earlier. With this motivation, we concentrate on automatically generating efficient test cases using UML state chart diagrams.

The organization of this chapter is as follows: Sect. 2 includes the background study, Sect. 3 describes related works, Sect. 4 shows the framework for generating test cases, and Sect. 6 comprises the conclusion and future work.

2 Background Study

To address the broad objective discussed in the previous section, some basic concept related to our work (synchronous testing) is elaborated [24]. Code-based testing is done after the code is created. In code-based testing, the tester tests each coverage as discussed above in Sect. 1. In model-based testing, it has all states; all transitions (AT); round trip paths (RTP); all transition pairs (ATP) with LN2, LN3, and LN4; full predicate; and all-paths coverage. The case study is shown in Fig. 2 as *ATM card validation*. We use Umbrello UML Modeller [25, 26] tool to draw the state chart diagrams. We have used EclEmma [27, 28] a free Java code coverage tool for Eclipse. It is used to show the percentage covered for the state and transition. In all state coverage criteria, each state must be covered at least once [29]. In all transition (AT), every transition is exercised at least once without any specific order. It covers all states, events, and actions [30]. This criteria cannot be avoided in testing.

Here, Fig. 2 is considered an example for state chart diagram. It shows the sequence of transitions and states for ATM card validation. It has six states: ATM idle, cardRead, Pin entry, verification, session next, and returning card. It exhibits the following valid state coverage:

1. (ATM idle, cardRead, Pin Entry)
2. (ATM idle, cardRead, Pin Entry, Verification, Session Next)
3. (ATM idle, cardRead, Pin Entry, Verification, Returning Card)

In this way, we use the coverage criteria AT, RTP, ATP, ATP with LN2, ATP with LN3, and ATP with LN4 to traverse the paths. We also generate test cases by considering the event sequences in the state machine.

AT: T (I) = (t1, t2, t3)/(create, card entry, reading card successfully); (t1, t2, t3, t4)/(create, card entry, reading card successfully, verify pin); (t1, t2, t3, t4, t5)/(create, card entry, reading card successfully, verify pin, abort); (t1, t2, t3, t4, t5, t6)/(create, card entry, reading card successfully, verify pin, abort, stopping pro-

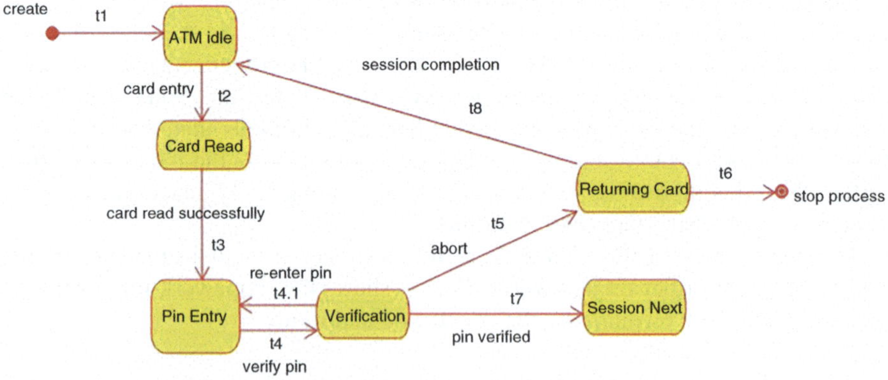

Fig. 2 State chart diagram for ATM card validation of class ATM card

cess); (t1, t2, t3, t4, t4.1)/(create, card entry, reading card successfully, verify pin, re-enter pin); (t1, t2, t3, t4, t7)/(create, card entry, reading card successfully, verifying pin, pin verified); (t1, t2, t3, t4, t5, t8)/(create, card entry, reading card successfully, verify pin, complete session). Test suite for AT, [T (I)] = 7.

RTP: (t1, t2, t3, t4, t5, t8, t1); (t1, t2, t3, t4, t4.1, t4, t5, t8, t1). Test suite for RTP, [T (II)] = 2.

ATP: T (III) = (t1, t2, t3, t4); (t1, t2, t3, t4); (t2, t2, t3, t4, t5, t8); (t1, t2, t3, t4, t5, t6). Test suite for ATP, [T (III)] = 3.

ATP with LN2: T (IV) = (t1, t2, t3) [one test case].

ATP with LN3: T (V) = (t1, t2, t3, t4) [one test case].

ATP with LN4: T (VI) = (t1, t2, t3, t4, t5); (t1, t2, t3, t4, t5, t8); (t1, t2, t3, t4, t5, t6); Test suite for ATP with LN4, [T (VI)] =3.

In mutation testing, bugs are intentionally inserted into the program, and mutants are created. A mutant is simply a faulty program. The tester aims is to determine the injected bugs that the test suite detected. The injected bugs represent the common faults that it may find in a real program. Choosing an efficient and effective test suite is one of the major tasks in software engineering. So, mutation testing plays an indispensable role in that purpose.

3 Related Work

We discuss the related works for model-based testing using state-based coverage criteria. The mutation score of FP is similar or higher compared to ATP. We select the AT, RTP, and ATP coverage criteria for this chapter. AT coverage found to be inadequate in the level of fault detection [31]. ATP coverage is stronger than AT and RTP with higher costs as it kills more mutants reported in paper [30]. Comparing AT, ATP, and RTP criteria, RTP was shown to be more cost-effective [32]. From the literature, it is seen that subsequent study said that for a weaker form of RTP coverage, which is not likely to be sufficient where significant numbers of faults remained undetected [31–33]. Then RTP was compared to random testing, which says using RTP 88% of the faults were detected. However, 69% faults were detected in random testing. Some typical faults are undetected in RTP testing [32]. Combining RTP with category-partition (CP) testing, the faults were detected with increasing cost. Subsequently, combining RTP with white box, the testing result shows better fault detection [34, 35]. A study shows that more than one transition tree can be generated by RTP strategy [32, 36]. Due to more transition tree generated, cost effectiveness can be increased. To improve the fault-detection effectiveness, Briand et al. [37] suggest some methods using RTP criterion. They investigated that with OCL guard conditions data flow analysis improves cost-effectiveness using various coverage criteria [37]. The results suggested that the generated transition tree has maximum fault-detection ability in data flow information [36, 37]. In a small, nonindustrial program, cruise control system [38], they found that a small percentage of the seeded faults was real and artificial faults were overwhelmingly reported [24, 31, 39]. There are studies for structural testing approaches for detecting realistic faults. Some studies are there which were executed in industrial settings [39, 40]. Of those studies, one study [39] did not mention about seeded faults. Another study [40] revealed the use of seeded mutants. Unlike testing sequential programs, testing concurrent programs require special types of coverage criteria. The concurrency coverage criteria handle a huge number of interleaving sequential test paths. The test paths are executed parallel. The rendezvous coverage criteria (RCC) is proposed by R. Yang and C.G. Chung [41]. The RCC include rendezvous node coverage (RN), rendezvous branch coverage (RB), rendezvous route coverage (RR), and concurrent route coverage (CR) [41]. Then further investigations are done for concurrency coverage criteria which is scalable and practical. The study evaluates the scalability of four concurrency coverage criteria (APESS, APESSnT, O-RESS, and NO-RESS). They introduce APESSnT, O-RESS, and NO-RESS coverage criteria. They further compare between the scalability of these concurrent coverage criteria with large numbers of processes, sequential test paths, and transitions.

4 Framework for Generating Test Cases

We propose a framework to generate test cases which is shown in a flowchart in Fig. 3.

The diagram clarifies the steps in this way. The state chart diagram is constructed using *StarUML* tool. Then the diagram is converted to an appropriate intermediate graph which we term as state chart intermediate graph (SCIG). Then coverage criteria are applied to the graph to generate possible testing paths. Next, test cases are generated from the testing path using different coverage criteria.

5 Case Study Implementation

Case Study 1

The state diagram of Fig. 4 has five states: initial, empty, holding, full, and final, respectively. The state, transition, and events are tested by traversing transition from one state to another in the state chart diagram.

We introduce an intermediate graph as state chart intermediate graph (SCIG) in our examples. An intermediate graph gives stepping information about the state machine, where the order of states and transitions is mentioned. Various intermediate graphs are previously used such as testing flow graph (TFG) [42], state transition graph (STG) [43], direct graph [44], flow graph (FG) [19], etc. where nodes and edges are mentioned in an order. Traversing those graphs, test cases are generated. TFG explicitly identifies flows of UML state chart diagrams and enhances for testing. From TFG test cases are generated using the testing criteria for the state and

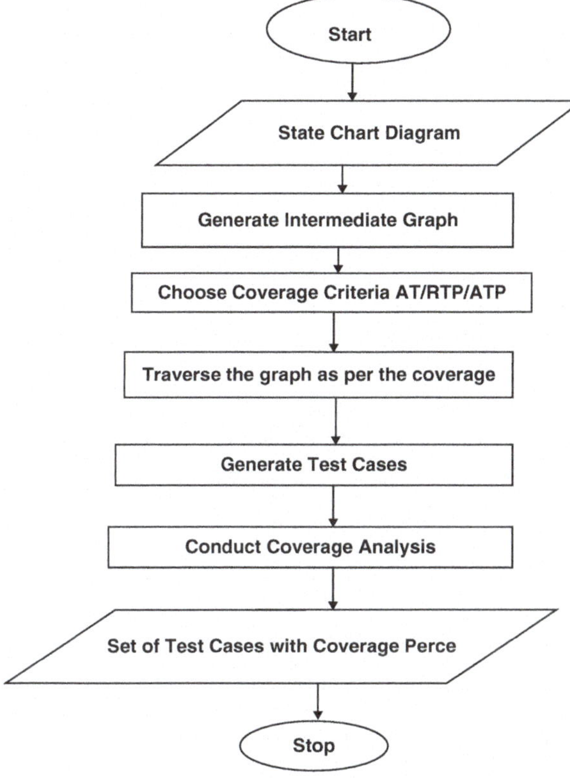

Fig. 3 Flow chart showing the test case generation

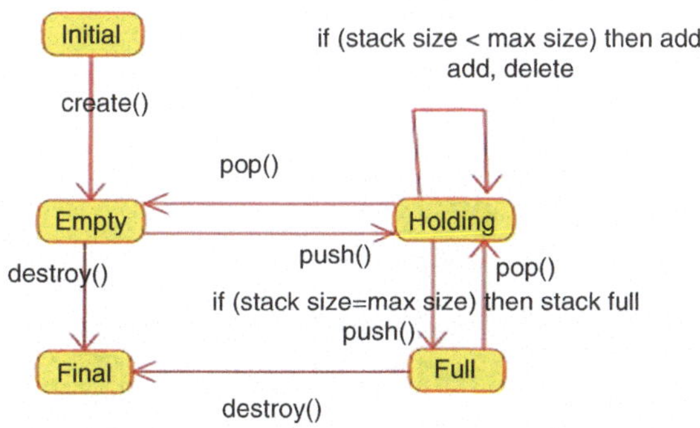

Fig. 4 State diagram of stack operation

transition of diagrams. SCIG of stack operation is shown in Fig. 5. SCIG is a directed graph that shows the order of states and transitions to be tested. The graph edges represent transitions, and the circles represent the states. Here, we assume that there is a unique node that corresponds to the start state. Other nodes represent the state change from the start node to the end node. The state diagram stack operation has eight transitions and five states. We develop Java-based XMI parser to analyze the XMI code of state chart diagram to generate SCIG. A snapshot of Java-based code of stack operation is shown in Fig. 6.

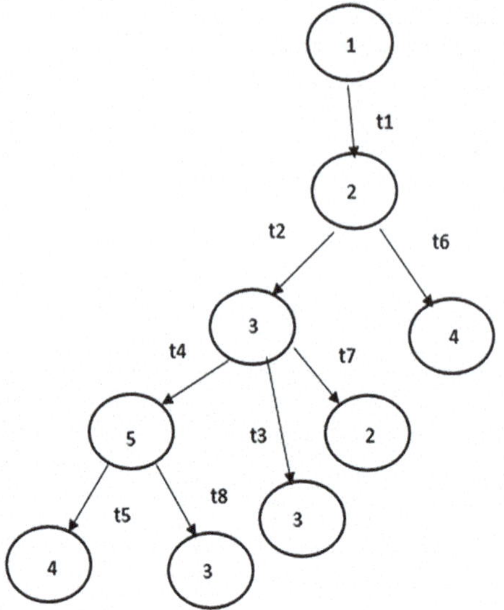

Fig. 5 SCIG of stack operation

```
package stack1;
public class stk
    {
    private int maxSize;
    private long[] stackArray;
    private static int top;
    public void create(int s)
    {
        maxSize=3;
        stackArray=new long[maxSize];
        top=-1;
    }
    public void push(long j){
        if(isFull())
        {
            System.out.println("Stack is Full");
            return;
        }
        stackArray[++top]=j;
    }
    public long pop(){

        if(isEmpty())
```

Problems @ Javadoc Declaration Console Coverage Cal

```
empty state
holding state
empty state
full state
holding state
holding state
20
stack destroyed
```

Fig. 6 Program implementation for stack operation

Case Study 2

Our next figure (Fig. 7) shows another case study, soft drink vending machine. We generate test transition sequences using AT, RTP, and ATP coverage criteria.

A state chart diagram of soft drink vending machine automation system object is shown in Fig. 7. The sequences of various events are taken into consideration to generate test cases. The object enters into various states working with different events. For example, covering the states and transitions, we find different test cases for the state model. Such as test case 1 (idle, showing option to select, displaying message, collecting money, dispersing drinks), test case 2 (idle, showing option to

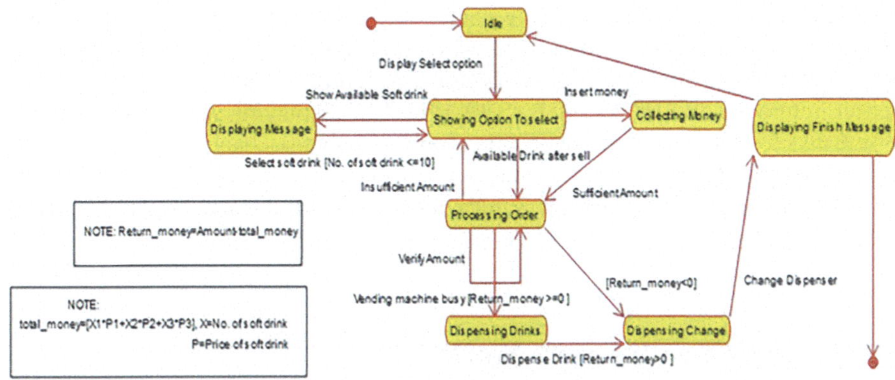

Fig. 7 A state chart diagram for the soft drink vending machine

select, displaying message, collecting money, dispersing drinks, dispersing change), test case 3 (idle, showing option to select, displaying message, collecting money, dispersing drinks, displaying finish), and some other test cases. Once select drink type event is triggered with the condition number of soft drinks <= 10), it goes to state *Displaying Message*. Then prices for different soft drinks are displayed. As the vending machine cannot deliver more than ten soft drinks, the condition X < = 10 is selected for each transaction. We calculate the $\text{Total}_{\text{money}} = (x1 \times p1 + x2 \times p2 + x\ 3 \times p3)$ making a category (cat = category) of soft drinks as X1 for cat 1 soft drink, X2 for cat 2 soft drink, and X3 for cat 3 soft drink. The variables P1, P2, and P3 are the soft drink prices for cat 1, cat 2, and cat 3 described earlier.

All Transition (AT)

AT is considered as being the minimum amount of coverage that one ought to accomplish in software testing. The test cases are as follows.

AT {TC1 = (t1, t2)/(displaying selected option, insert money), TC2 = (t1, t3)/(displaying selected option, show available soft drink), TC3 = (t1, t3, t4)/(displaying select option, show available soft drink, select soft drinks [no. of drinks <=10]), TC4 = (t1, t3, t4, t5)/(displaying select option, show available soft drink, select soft drinks [No. of drinks <=10], available drinks after sell), TC5 = (t1, t2, t6)/(displaying selected option, insert money, sufficient amount), TC6 = (t1, t2, t6, t7)/(displaying selected option, insert money, amount, verify amount, insufficient amount), and many more.

Round Trip Path (RTP)

RTP describes the beginning state as well as the end state and it is the same for all paths in the state machine. A breadth- or depth-first search algorithm is used to construct a transition tree. The repetition of the same node is the final node or end node in the transition tree.

RTP {TC12 = (t1, t3, t4)/(displaying select option, show available soft drink, select soft drinks [No. of drinks <=10]), TC13 = (t1, t5, t8, t10, t11, t12)/(displaying select option, available drinks for sell, verify amount, return_money > = 0, change dispenser, return to idle), TC14 = (t1, t5, t9, t11, t12)/(displaying select option, available drinks for sell, return_money > = 0, return_money >0, change dispenser, return to idle), TC15 = (t1, t2, t6, t8, t10, t11, t12)/(displaying select option, insert money, sufficient amount, verify amount, return_money > = 0, change dispenser, return to idle).

6 Conclusion with Future Work

We focused on generating test cases using state chart diagrams. We considered various coverage criteria to traverse the path in the state machine. This method assisted us to detect various state-based faults. The common state-based faults we found are missing transitions, missing states, incorrect events, etc. Here we gave emphasis on various coverage criteria using different state models. Although SBT is not a new research area, but this chapter presents a different aspect to evaluate state-based testing. We observed different test cases by implementing various state-based coverage criteria in various case studies. The researchers use various intermediate graphs such as testing flow graph (TFG), flow graph (FG), state transition graph (STG), and abstract syntax tree in different papers. We proposed an appropriate SCIG for the state model to avoid the state explosion problem. In future work, we will be interested in test suite reduction using the concept of fuzzy logic.

References

1. Antoniol, G., Briand, L. C., Di Penta, M., & Labiche, Y. (2002). A case study using the round-trip strategy for state-based class testing. In *Proceedings of the13th international symposium on software reliability engineering* (pp. 269–279). https://doi.org/10.1109/ISSRE.2002.1173268
2. Mishra, D. B., Acharya, A. A., & Acharya, S. (2020). White box testing using genetic algorithm—An extensive study. In *A journey towards bio-inspired techniques in software engineering* (Vol. 185, p. 167). Springer.
3. Agrawal, H., DeMillo, R., Hathaway, R., Hsu, W., Krauser, E., & Spafford, E. (1989). *Design of mutant operators for the C programming language* (Technical report SERC-TR-41-P). Software Engineering Research Center, Department of Computer Science, Purdue University.
4. Holt, N. E., Briand, L. C., & Torkar, R. (2014). Empirical evaluations on the cost-effectiveness of statebased testing: An industrial case study. *Information and Software Technology, 56*(8), 890–910. https://doi.org/10.1016/j.infsof.2014.02.011
5. Kaner, C., & Padmanabhan, S. (2007). Practice and transfer of learning in the teaching of software testing. In *20th Conference on Software Engineering Education & Training (CSEET'07)* (pp. 157–166). IEEE.
6. Mishra, D. B., Mishra, R., Das, K. N., & Acharya, A. A. (2019). Test case generation and optimization for critical path testing using genetic algorithm. In *Soft computing for problem solving* (pp. 67–80). Springer.

7. Briand, L., & Labiche, Y. (2001). A UML-based approach to system testing. In *International conference on the unified modeling language* (pp. 194–208). Springer.
8. Mishra, D. B., Mishra, R., Das, K. N., & Acharya, A. A. (2017). A systematic review of software testing using evolutionary techniques. In *Proceedings of sixth international conference on soft computing for problem solving* (pp. 174–184). Springer.
9. Farchi, E., Hartman, A., & Pinter, S. S. (2002). Using a model-based test generator to test for standard conformance. *IBM Systems Journal, 41*(1), 89–110.
10. Neto, A. D., Subramanyan, R., Vieira, M., Travassos, G. H., & Shull, F. (2008). Improving evidence about software technologies: A look at model-based testing. *IEEE Software, 25*(3), 10–13.
11. Dias-Neto, A. C., & Travassos, G. H. (2010). A picture from the model-based testing area: Concepts, techniques, and challenges. In *Advances in computers* (Vol. 80, pp. 45–120). Elsevier.
12. Shafique, M., & Labiche, Y. (2010). *A systematic review of model based testing tool support* (Technical report SCE-10-04). Carleton University.
13. Nidhra, S., & Dondeti, J. (2012). Black box and white box testing techniques-a literature review. *International Journal of Embedded Systems and Applications (IJESA), 2*(2), 29–50.
14. Bohme, M., Pham, V. T., & Roychoudhury, A. (2017). Coverage-based greybox fuzzing as Markov chain. *IEEE Transactions on Software Engineering.*
15. Binder, R. V. (2000). *Testing object-oriented systems: Models, patterns, and tools*. Addison-Wesley Professional.
16. Broy, M., Jonsson, B., Katoen, J. P., Leucker, M., & Pretschner, A. (2005). *Model-based testing of reactive systems. Advanced lectures: Outcome of a research seminar*. Springer. https://doi.org/10.1007/b137241
17. Utting, M., Pretschner, A., & Legeard, B. (2012). A taxonomy of model-based testing approaches. *Software Testing, Verification and Reliability, 22*(5), 297–312.
18. El-Far, I. K., & Whittaker, J. A. (2001). Model-based software testing. In *Encyclopedia of software engineering.*
19. El-Fakih, K., Simao, A., Jadoon, N., & Maldonado, J. C. (2017). An assessment of extended finite state machine test selection criteria. *Journal of Systems and Software, 123*, 106–118.
20. Chow, T. S. (1978). Testing software design modeled by finite-state machines. *IEEE Transactions on Software Engineering, SE-4*(3), 178–187. https://doi.org/10.1109/TSE.1978.231496
21. Turner, C. D., & Robson, D. J. (1993). The state-based testing of object-oriented programs. In *Proceedings software maintenance CSM-93 conference on IEEE* (pp. 302–310).
22. Utting, M., Legeard, B., Bouquet, F., Fourneret, E., Peureux, F., & Vernotte, A. (2016). Recent advances in model-based testing. *Advances in Computers, 101*, 53–20.
23. Pradhan, S., & Ray, M. (2021). Asynchronous testing in web applications. In *Intelligent and cloud computing* (pp. 355–361). Springer.
24. Abdurazik, A., Ammann, P., Ding, W., & Offutt, J. (2000). Evaluation of three specification-based testing criteria. In *Proceedings sixth IEEE international conference on engineering of complex computer systems. ICECCS 2000* (pp. 179–187). IEEE.
25. Toth, K. (2006). Experiences with open source software engineering tools. *IEEE Software, 23*(6), 44–52.
26. Safdar, S. A., Iqbal, M. Z., & Khan, M. U. (2015). Empirical evaluation of UML modeling tools–a controlled experiment. In *European conference on modelling foundations and applications* (pp. 33–44). Springer.
27. Arcuri, A., Fraser, G., & Galeotti, J. P. (2014). Automated unit test generation for classes with environment dependencies. In *Proceedings of the 29th ACM/IEEE international conference on automated software engineering* (pp. 79–90).
28. Polo, M., Reales, P., Piattini, M., & Ebert, C. (2013). Test automation. *IEEE Software, 30*(1), 84–89.
29. Utting, M., & Legeard, B. (2010). *Practical model-based testing: A tools approach*. Elsevier.

30. Holt, N. E., Torkar, R., Briand, L., & Hansen, K. (2012). State-based testing: Industrial evaluation of the cost-effectiveness of round-trip path and sneak-path strategies. In *2012 IEEE 23rd international symposium on software reliability engineering* (pp. 321–330). IEEE.
31. Briand, L. C., Labiche, Y., & Wang, Y. (2004). Using simulation to empirically investigate test coverage criteria based on statechart. In *Proceedings of the 26th international conference on in software engineering* (pp. 86–95). ICSE IEEE.
32. Briand, L.C., Labiche, Y., & Lin, Q. (2005). Improving statechart testing criteria using data flow information. In *16th IEEE International Symposium on Software Reliability Engineering (ISSRE'05)* (p. 10). IEEE.
33. G. Antoniol, L. Briand, M. Di Penta, Y. Labiche. (2002). A case study using the round-trip strategy for state-based class testing. In *Proceedings of the 13th international symposium on software, reliability engineering (ISSRE'02).*
34. S. Mouchawrab, L. Briand, Y. Labiche. (2007). Assessing, comparing, and combining statechart-based testing and structural testing: An experiment. In *First international symposium on empirical software engineering and measurement, ESEM 2007.*
35. Mouchawrab, S., Briand, L., Labiche, Y., & Di Penta, M. (2011). Assessing, comparing, and combining state machine-based testing and structural testing: A series of experiments. *IEEE Transactions on Software Engineering, 37*(2), 161–187.
36. Pradhan, S., Ray, M., & Swain, S. K. (2019). *Transition coverage based test case generation from state chart diagram.* Journal of King Saud University-Computer and Information Sciences.
37. Briand, L., Labiche, Y., & Lin, Q. (2010). Improving the coverage criteria of UML state machines using data flow analysis. *Software Testing, Verification and Reliability, 20*(3), 177–207.
38. Gomaa, H. (2006). Designing concurrent, distributed, and real-time applications with UML. In *Proceedings of the 28th international conference on software engineering* (pp. 1059–1060).
39. Bogdanov, K., & Holcombe, M. (2001). State chart testing method for aircraft control systems. *Software Testing, Verification and Reliability, 11*(1), 39–54.
40. Chevalley, P., & Thévenod-Fosse, P. (2001). An empirical evaluation of statistical testing designed from UML state diagrams: The flight guidance system case study. In *Proceedings 12th international symposium on software reliability engineering* (pp. 254–263). IEEE.
41. Yang, R.D., & Chung, C.G. (1990). A path analysis approach to concurrent program testing. In *Ninth annual international Phoenix conference on computers and communications. 1990 conference proceedings* (pp. 425–432). IEEE.
42. Kansomkeat, S., & Rivepiboon, W. (2003). Automated-generating test case using UML statechart diagrams. In *Proceedings of the 2003 annual research conference of the South African institute of computer scientists and information technologists on enablement through technology* (pp. 296–300).
43. Swain, R., Panthi, V., Behera, P. K., & Mohapatra, D. P. (2012). Automatic test case generation from UML state chart diagram. *International Journal of Computer Applications, 42*(7), 26–36.
44. Belli, F., Budnik, C. J., Hollmann, A., Tuglular, T., & Wong, W. E. (2016). Model-based mutation testing approach and case studies. *Science of Computer Programming, 120*, 25–48.

A Heuristic-Based Test Case Prioritization Algorithm Using Static Metrics

Daniel Getachew, Sudhir Kumar Mohapatra, and Subhasish Mohanty

1 Introduction

As computer hardware becomes cheaper, the focus transfers to software systems. Large software systems may be more complex than the hardware used to run them, so there is a great demand for best practices and engineering processes that can be applied to software development. Software engineering is a thickly folded and jumbled concept for those who want to have just a simple two-sentence definition about it. According to [1], it is a systematic implementation of technological and scientific knowledge, approach, and know-how to the design, implementation, testing, and documentation of software. Also as defined by [2], software engineering is a disciplined, quantifiable approach to the operation, development, and maintenance of software. Software testing is among the major phases of software engineering to verify and validate the quality of software. It is a technique of executing a program or system with the intent of finding defects which help us to confirm the product has been manufactured by programmers as quality product. Also it assures that the manufactured product is working as indicated on the requirement specification and achieves user satisfaction.

Software testing took a lion's share to develop a test case prioritization to boost the developer's confidence in which the program is developed as intended to do so. The main intention of software testing is to find faults in the program [3]. It can be considered as verifying and validating. The verification process could be considered

D. Getachew
Addis Ababa Science and Technology University, Addis Ababa, Ethiopia

S. K. Mohapatra (✉)
Faculty of Emerging Technologies, Sri Sri University, Cuttack, Odisha, India
e-mail: sudhir.mohapatra@srisriuniversity.edu.in

S. Mohanty
G.I.E.T University, Gunupur, Odisha, India

M. Khari et al. (eds.), *Optimization of Automated Software Testing Using Meta-Heuristic Techniques*, EAI/Springer Innovations in Communication and Computing, https://doi.org/10.1007/978-3-031-07297-0_4

white-box testing, and it also answers the question "are we building the software, right?" Also, the validation process addressed the question "are we building the right question?," and it is considered as black box testing [4].

Moreover, throughout the development of any software, there is always maintenance and modification on any existing stage of the software version at production. The changes made in that software may lead to the improvement of software features and fixing bugs or break software, that is, introducing new bugs into the software. Such breaking changes are called regression. So for the development team to make sure that the newly added modification does not cause any break on the software, they have to perform regression testing.

Regression testing is rerunning functional and nonfunctional tests to ensure that previously developed and tested software still performs fine after a change. As regression test suites tend to grow with each found defect, test automation is frequently involved. Sometimes a change impact analysis is performed to determine an appropriate subset of tests. Using software testing frameworks helps us to make an effective test on our software.

2 Related Work

Regression test selection (RST), test suite minimization (TSM), and test case prioritization (TCP) are leading regression testing strategies. The retest-all strategy holds well when the test suite is small. However, as the size of test suite increases, an ordering mechanism becomes necessary [5]. Test case prioritization (TCP) attempts to find an optimal execution order of test cases to get the maximum rate of fault detection. TCP has been an active field of research in software engineering for more than two decades, and myriad techniques have been proposed for performing it [6].

The TCP problem has been addressed in several ways by many researchers [7]. Dario et al. [8] noticed that area under coverage (AUC) metrics represent a bidimensional version of the hyper-volume metric, which is widely used in many-objective optimization. Thus, they propose a hyper-volume base genetic algorithm, namely, HGA, to solve the test case prioritization problem when using multiple test case prioritization criteria. Also, an empirical study which they have conducted concerning five state-of-the-art techniques (viz., additional greedy, multi-objective evolutionary algorithm based on decomposition, non-dominated sorting genetic algorithm II, generalized differential evolution 3, and a genetic algorithm based on an AUC metric shows that: (i) HGA is more cost-efficient, (ii) HGA improves the efficiency of test case prioritization, and (iii) HGA has stronger selective pressure when dealing with more than three criteria. Muhammed et al. [9] also have proposed and experiment using a firefly algorithm with fitness function defined using a similarity distance model. And the result shows that firefly obtained highest average percentage of faults detected (APFD) scores compared to other prioritization algorithms (particle swarm optimization (PSO), local beam search (LBS), greedy, and genetic algorithm (GA)), and also their experiment shows that firefly algorithm slightly outperforms LBS in terms of execution time.

Faiza et al. [10] research TCP optimization using a mutation testing-based prioritization technique. In this approach, they seed different faults in the original program to create multiple mutated copies of the program, and the test case that detects maximum number of faults gets highest priority. Their proposed technique shows an improvement in the rate of fault detection of test suites. Hence, most test case prioritization techniques are developed using some coverage criteria, and that makes this technique an exceptional white box prioritization technique. In addition to this Lei et al. [11] experiment clustering approach combining fault prediction which results that it can improve the effectiveness of test case prioritization. Wenhao et al. [12] developed a TCP technique based on method invocation relationship and program changes. By this approach, they combine method coverage information and estimated risk value of each program method. As a result, the prioritization problem got reduced to an integer linear programming problem, and this shows that their algorithm is more effective than some well-studied TCP techniques.

Paruchuri et al. [13] examine that incorporating requirements in testing phase could help a lot in finding the faults and errors easily. Although there are many prioritizing techniques with using source code information, the effectiveness is not up to the mark. The percentage of productivity has been increased to 80% by using requirement information in prioritizing the test cases. Dipesh et al. [14] have also discovered that rule mining and multi-objective search (named as REMAP) technique for dynamic test case prioritization can significantly outperform the other approaches. Moreover, Yijie et al. [15] have experiment based on GUI software features such as event handler and function call graph using two centrality measures (degree centrality and betweenness centrality and found that the combination of centrality method and existing conventional method has a high potential for improving prioritization effect.

Qi Luo et al. [16] investigated the performance of static and dynamic prioritization techniques on modern software systems by selecting four static and state of research dynamic techniques. As per experimental results on two evaluation metrics (APFD and APFDc), it showed that these metrics incline to correlate test-class granularity, but this correlation does not hold at test-method granularity. Moreover, static prioritization techniques better perform when evaluation is using APFDc. When TCP techniques are applied on large-scale programs, they perform better, yet the size of the program does not affect the performance measure between techniques while comparing them. The performance result between TCP techniques does not be impacted by software evolution. Regarding the effect of mutants, both the number and type of mutant applied do not affect measures at TCP effectiveness under experimental settings. The illustration of similarity analysis shows that highly prioritized test cases tend to uncover unrelated faults.

Even though there are a bunch of researches on optimal technique, it seems that it is potentially promising for the improvement of TCP, and some studies indicated that optimal technique is not that beneficial in terms of fault detection rate on execution time [17]. As regression testing is not a one-time activity, prioritizing test cases using historical information about test case performance record or historical fault information was also explored by researchers [18].

Coverage aware prioritization techniques aim to maximize coverage of program elements (statement/branch/methods, etc.) by a test case. They require detailed knowledge of source code. Rothermelet al. [19] showed that better coverage yields better fault detection rate. In addition to this, prioritization using dynamic coverage information collected from early versions performs better than those using static coverage information [20]. The limitation of these works was that it considered all faults of the same severity and did not incorporate the cost factor. Also [10] testifies that mutation-based prioritization technique addressed the major drawback of branch coverage-based prioritization technique which is assigning lower priority for the test cases that expose maximum number of faults; hence, in mutation-based TCP technique, higher priority is assigned to those test cases.

3 The Proposed Prioritization Algorithm (StatPriori)

```
Input: Program with initial test suite TS, Cyclomatic Complexity C, Halsted's Metrics Volume as Vol, and Halsted's Metrics Vocabulary as Voc.
Output: Prioritized test cases
for(TS ϵ P) do
    for(T_a ϵ TS) do
        for(a ϵ n)
            i←a
            k←a+1
            for(k ϵ a) do
                if(C_i < C_k)
                    i←k
                else if(C_i == C_k)
                    if(Vol_i < Vol_k)
                        i←k
                    else if(Vol_i == Vol_k)
                        if(Voc_i < Voc_k)
                            i←k
            Swap(T_a, T_i)
            TS'← TS
        return TS'
```

The above written algorithm takes the program test suite with corresponding metrics value as initial input, and it is expected to return the prioritized algorithm which is the main objective of this research. After taking its initial input, it checks the value of the metric of every test case in the test suit to assign priority. The compression is based on three metrics which are more concerned about faults,

- McCabe Cyclomatic complexity concerns that the highest the Cyclomatic complexity the highest there is a probability of faults in that segment. According to prior researches like [21, 22], it claims that there is a correlation between the complexity of the system and the number of faults.
- Halstead's volume metrics tells that the highest the volume of the program the highest the count of the number of mental comparisons required to generate a program [23], which could lead the implementation of the algorithm being faulty and that need early execution.
- Also, Halstead's vocabulary of a program segment tells about how many tokens could be in that segment of the program. The higher the vocabulary the higher having token diversity, which could lead faulty tokens to be included (Fig. 1).

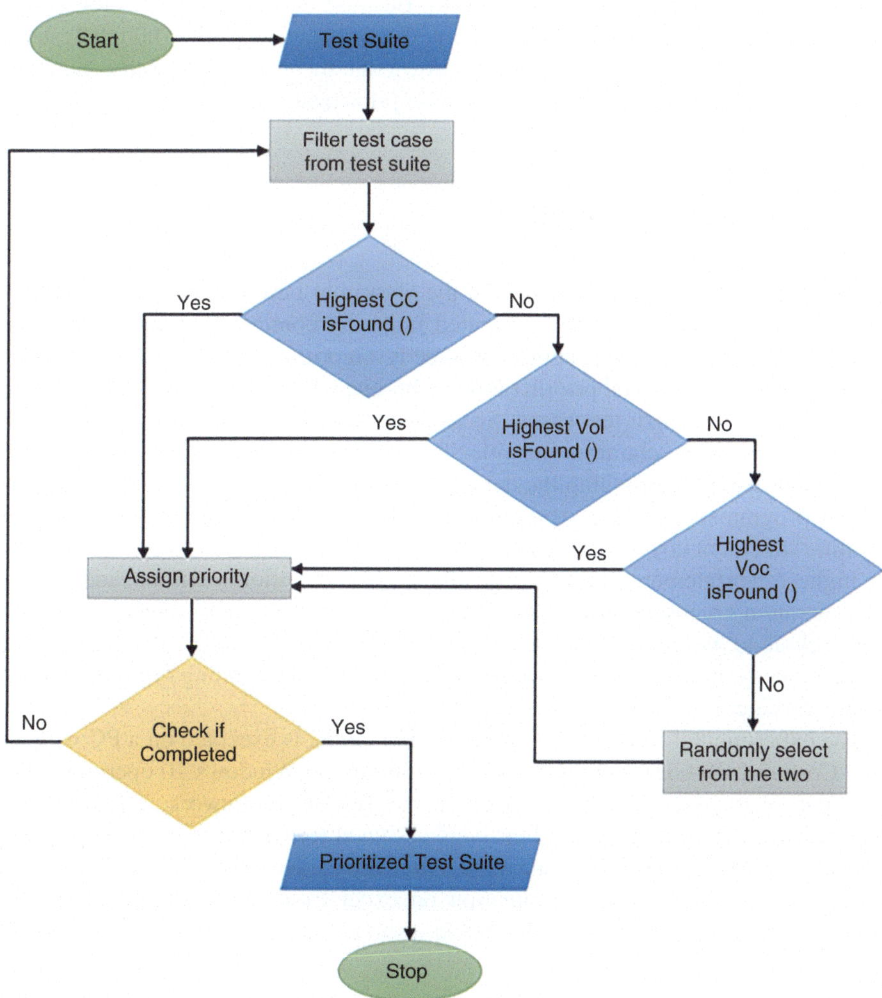

Fig. 1 Flowchart of the algorithm

Having these three metrics the algorithm compares the test cases with the condition when two test cases are compared, and the test case having the highest Cyclomatic complexity will get the highest priority, but if both have equal Cyclomatic complexity, the algorithm will compare their volume, and the one which has the highest volume will get the highest priority again if tie condition occurs for the third time, and it will check the vocabulary and gives the highest priority to the one having the highest value of vocabulary. After these three metrics comparison if there is a probability of tie condition happening between two test cases, it will give priority based on their previous position. Generally, the algorithm works by sorting based on the metrics by swapping their positions.

Nonetheless, TCP techniques without coverage information have shown promising results, and it is still not that popular. Collecting coverage information is a costly effort in terms of time, effort, and cost [18]. Because of this, many researchers are trying to avoid this step. In all the literature, the authors use JUnit. Though TestNG is released long back, to the best of our knowledge there is no research conducted on TestNG specifically in the area of test case prioritization.

4 Experiment and Result

For each subject program, we executed the algorithm based on the test suite pool based on the test cases and the generated faults by comprising three benchmark algorithms, namely, random approach, greedy algorithm based on the statement coverage, and method-total prioritization technique which we use method coverage information to prioritize test cases. So, we had fair comparison because it is our exception case as we claim on our title. The random approach is employed in our experiment just by rearranging the default order of the test cases in our test suite of subject programs. So while arranging the called statements of the test case, we examine statement coverage of each test case, and we order the test cases based on the highest the statement coverage value the highest the priority will be, and by that we generate a greedy algorithm-based prioritized test suite pool. For the method-total technique, we record the trace of test case method coverage after execution, like the greedy approach we rearrange the test cases in descending order of method coverage value.

The experimental setup for the implemented model is executed on a PC with an Intel Core i5 2.40 GHz and 8 GB memory running on Windows 10 operating system, and all the execution takes place on the TestNG framework. The priorities obtained from these three algorithms were assigned for each test case using @Test annotation using the "priority" attribute. After the execution of each test case, we record the test case—fault combination on excel from the report generated by TestNG so that we can calculate the APFD result of each algorithm later.

4.1 Result

Based on the size of the test suite pool of the subject program, we record different sizes of test cases for the percentage of test case fraction on 15 intervals for each subject program. The number of test cases in each interval is defined based on the size of test cases in the test suite pool ranging from 5 to 15. The result recorded for each subject program by applying the algorithms is recorded as follows (Tables 1 and 2).

Table 1 Percentage of faults detected per percentage of test suite fraction of Store Project and Grade Book

	Store project				Grade book			
PTSF	Random	GA	MTT	StatPriori	Random	GA	MTT	StatPriori
0	0	0	0	0	0	0	0	0
0.083	7.55	9.26	13.2	15.09	6.25	9.375	15.62	18.75
0.166	15.09	16.98	28.3	26.42	12.5	15.62	25	31.25
0.249	20.75	24.53	33.96	30.19	15.62	31.25	34.38	50
0.332	30.2	32.08	37.73	37.74	21.87	40.62	46.88	53.125
0.415	35.84	39.62	41.5	47.17	31.25	40.62	56.25	68.75
0.498	41.51	49.06	56.6	60.38	34.38	50	65.63	68.75
0.581	52.83	58.49	81.13	83.02	43.75	56.25	81.25	87.5
0.664	64.15	69.81	88.68	92.45	53.12	65.63	96.88	93.75
0.747	81.13	84.9	98.11	100	65.63	78.12	100	100
0.83	88.67	96.23	100	100	84.37	90.62	100	100
0.913	98.11	100	100	100	96.87	96.87	100	100
1	100	100	100	100	100	100	100	100

Table 2 Percentage of faults detected per percentage of test suite fraction of Sudoku Program and STACK

	Sudoku program				STACK			
PTSF	Random	GA	MTT	StatPriori	Random	GA	MTT	StatPriori
0	0	0	0	0	0	0	0	0
0.083	5.13	7.7	12.83	12.83	6.98	11.63	18.61	27.91
0.166	7.7	15.4	23.08	20.52	13.96	16.28	25.58	37.21
0.249	20.52	25.65	30.77	38.46	18.61	20.94	34.88	39.53
0.332	25.65	35.9	43.59	48.72	30.24	30.24	44.19	51.16
0.415	35.9	48.72	61.54	66.67	41.87	30.24	53.49	58.14
0.498	43.59	53.85	61.54	76.93	48.84	48.84	62.8	62.79
0.581	53.85	64.1	76.93	84.62	58.14	60.47	69.77	76.74
0.664	58.98	66.67	87.18	89.74	74.42	76.75	88.37	88.37
0.747	71.8	82.06	97.44	100	83.73	90.7	95.35	93.02
0.83	87.18	94.88	100	100	95.35	95.35	100	97.67
0.913	92.31	100	100	100	100	100	100	100
1	100	100	100	100	100	100	100	100

The above two tables showed the result recorded while executing three of benchmark prioritization techniques and StatPriori on the subject programs. PTSF stands for percentage of test suite fraction. The second column labeled with random holds the percentage of faults detected while applying random prioritization technique, GA represents greed algorithm, the column labeled with GA presents the percentage of faults detected while applying greedy prioritization approach using statement coverage, MTT stands for method-total technique, the column presents the faults detected while applying method-total technique, and the column labeled with StatPriori presents the result obtained while applying StatPriori algorithm. The following section discusses each subject program and the TCP techniques performance in each subject program.

4.1.1 Store Project

On the result table, the performance of the TCP techniques when applied on Store Project test suite pool is recorded with 11 test case execution interval. As we can see from Table 3 after executing 22–33%, less than 25% of faults were detected, while the greedy algorithm applied on 13 out of 53 faults was detected, and the fault detection rate of StatPriori was slightly higher than 30%. Surprisingly, MTT detects 18 out of 53 after executing 33 test cases, and that's 33.96% of the total faults seeded. However, after executing 50% of test cases, StatPriori reaches 60% of fault detection, and yet random and GA were less than 50% detection rate, and even MTT approach was able to detect 56.6% of test cases. MTT requires at most 110 out of 132 test cases, StatPriori requires almost 99 test cases to be executed, and the random and GA approach require almost 132 and 121 test cases to be executed to detect all the faults. The APFD result is examined on.

The APFD result obtained from the implementation of Store Project showed that StatPriori prioritization technique outperforms all the three benchmark prioritization techniques. The performance gap between MTT and StatPriori when compared with random and greedy algorithm is more than 8%. The performance of StatPriori is examined as 62.02%, yet the list performance was 49.02% resulted by random prioritization technique (Fig. 2).

4.1.2 Grade Book

Grade Book was the largest subject program based on the line of codes that contain and test pool size. The execution result presented in Table 4 was recorded at 15 test case execution so that each row of the table contains 15 test cases. As we can see

Table 3 APFD result of obtained at Store Project execution

	Random (%)	GA (%)	MTT (%)	StatPriori (%)
APFD	49.02	52.77	60.92	62.02

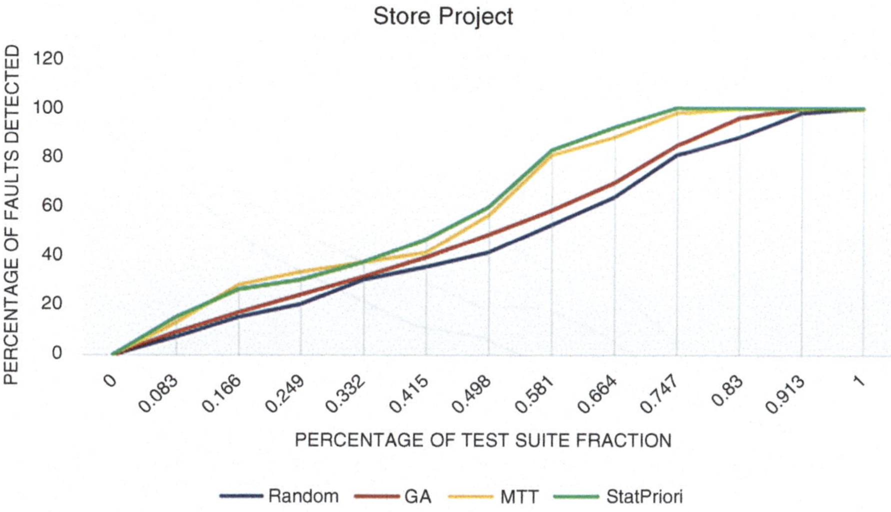

Fig. 2 APFD chart of the Store Project execution

Table 4 APFD result of obtained at Grade Book execution

	Random (%)	GA (%)	MTT (%)	StatPriori (%)
APFD	43.18	52.26	64.46	68.61

from the execution results, MTT detects 25% of faults after executing almost 30 test cases, and yet the random and greedy approach detects four and five faults, respectively. StatPriori detects 50% of faults seeded by executing 30% of test cases. However, greedy algorithm needs 50% to reach this level of fault detection. The random and greedy approach needs 100% of test cases to be executed, yet both MTT and StatPriori need 75% of test cases to be executed to detect all the test cases. The following table presents the APFD result of the execution.

In this subject program, the performance of StatPriori is higher than Store Project, and the random prioritization approach performs lower than Store Project. The percentage gap between MTT and StatPriori exceeds more than 4%, which MTT and StatPriori showed 64.46% and 68.61%, respectively (Fig. 3).

4.1.3 Sudoku Program

As presented in Table 5, the performance of the algorithms is recorded while applying them on the Sudoku Program. The result is recorded at six test case execution interval. At the very beginning of the executions, MTT shows better performance by detecting 9 out of 39 faults at the rate of 23.08%. Furthermore, the performance of StatPriori, GA, and random prioritization were 20.52, 15.4, and 7.7, respectively. The performance rate of MTT gets stuck at 61.54% when it reaches 41.66% till it

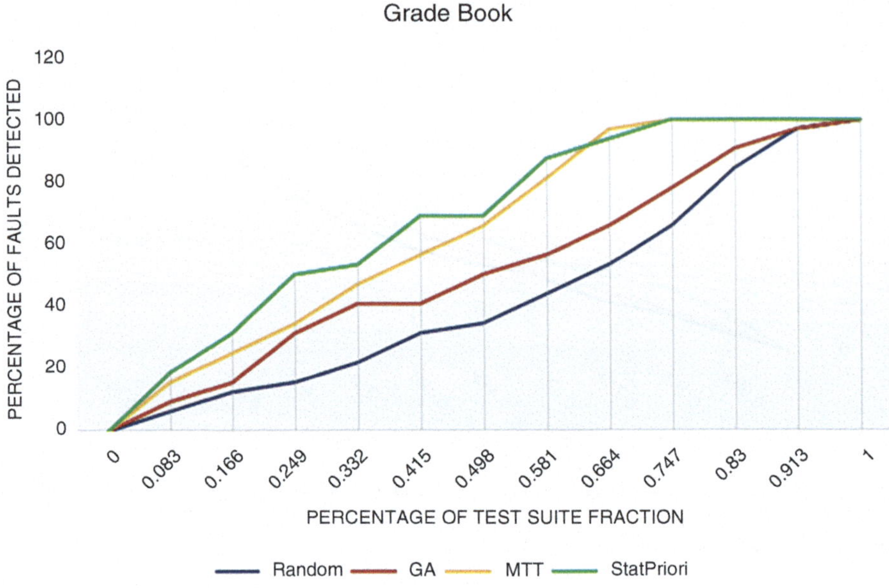

Fig. 3 APFD chart of the Grade Book execution

Table 5 APFD result of obtained at Sudoku Program execution

	Random (%)	GA (%)	MTT (%)	StatPriori (%)
APFD	46.25	53.92	62.22	65.84

reaches 50%. However, StatPriori requires almost 54 test cases to complete the whole execution, but MTT requires six additional test cases to finish its execution. Moreover the performance of random and greedy algorithms was better than earlier execution, but it was at a lower performance level of the current execution.

The performance of StatPriori recorded on this subject program is less than Grade Book program and better than Store Project, and the vice versa is true for random prioritization approach. Surprisingly StatPriori performs better on this program too, and the performance rate was 65.84%. Regardless of the performance at the beginning of the execution, MTT exhibits the next better performance rate at 62.22% (Fig. 4).

4.1.4 STACK

The recorded result with interval of five test case execution while applying the algorithms on STACK program shows that random and greedy algorithm prioritization approach performs the same 30.24% detection rate while executing 20 of 60 test cases, at the same execution level StatPriori achieves more than 50% of fault detection. Also after executing 40 test cases, MTT and StatPriori reach at same detection

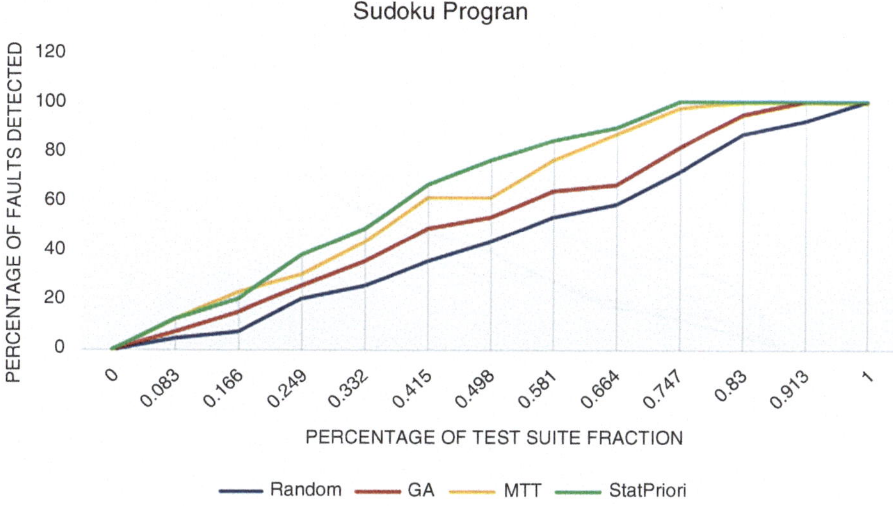

Fig. 4 APFD chart of the Sudoku Program execution

Table 6 APFD result of obtained at STACK execution

	Random (%)	GA (%)	MTT (%)	StatPriori (%)
APFD	52.03	52.80	62.07	65.35

performance rate of 88.37%. Finally, three of the algorithms (random, greedy, and StatPriori) required at most 55 test cases out of 60 to detect all of the faults in the test suite pool. However, MTT requires only 50 test cases to reach 100% fault (Table 6).

Fortunately, the execution completion of MTT was earlier than all of the applied algorithms, yet the overall performance was not as efficient as StatPriori. MTT achieves 62.07% of APFD result, but StatPriori achieves 65.35%. The performance of random and GA was very close to each other, exhibiting 52.03% and 52.80%, respectively (Fig. 5).

To have a quick recap of the above results, in each experiment the performance of the algorithms was examined, and it is found that the algorithms can be considered from the lower to higher as from random ordering to our developed algorithm (StatPriori) on each subject program but not compared to one another. On all of the four subject program, the experiment shows that the fault detection rate is increased and with minimum of 1.1% to maximum of 25.43% on Store Project and Grade Book, respectively. The minimum performance of random prioritization algorithm and the maximum performance of StatPriori were exhibited on Grade Book execution. Based on the APFD result obtained from the experiment, the comparison to drive any conclusion the comparison must not be between subject programs, and if it has to be between algorithms, it should be the performance of these algorithms applied to the same subject programs. This is not only because each subject program

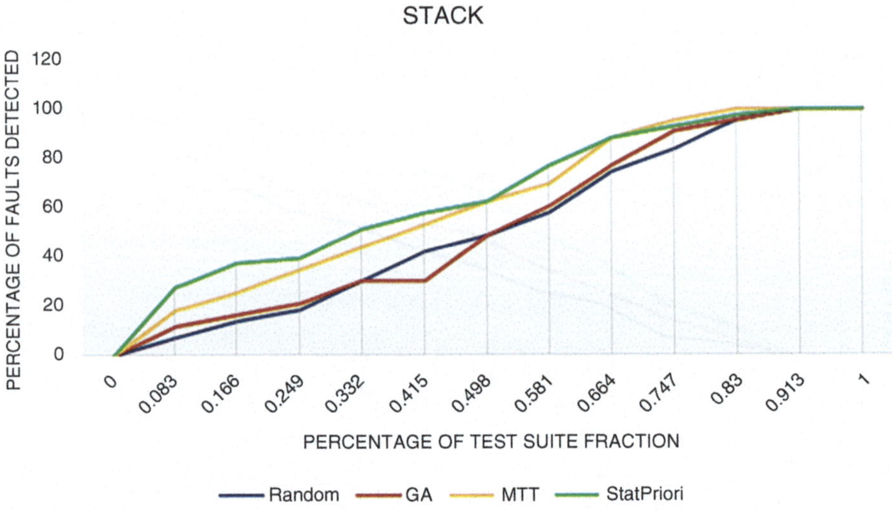

Fig. 5 APFD chart of the STACK execution

Table 7 Overall APFD result of each program on each algorithm

	Random (%)	GA (%)	MTT (%)	StatPriori (%)
Store Project	49.02	52.77	60.92	62.02
Grade Book	43.18	52.26	64.46	68.61
Sudoku Program	46.25	53.92	62.22	65.84
STACK	52.03	52.80	62.07	65.35
Minimum	43.18	52.26	60.92	62.02
Maximum	52.03	53.92	64.46	68.61
Average	47.62	52.94	62.42	65.46

has a different test suite pool size but also because the size of the subject programs is different too (Table 7).

5 Conclusion

The developed prioritization technique gives an order based on the comparison between these static metrics which the test case having the highest Cyclomatic complexity gets the highest priority, and if tie condition happens, the algorithm will compare the Halstead volume metrics of the source code, and the one which has the highest volume will get the highest priority again, and if another tie condition occurs, then the algorithm goes to compare the Halstead vocabulary and the test case which has highest vocabulary value will get the highest priority. Using multiple fitness object to prioritize test cases will have a strong tie-breaking capability and prioritize test cases; in this research using three fitness objectives to prioritize test cases makes the developed algorithm perform better on breaking tie conditions. A test case prioritization technique developed with static software metrics showed better performance on the experiment conducted on four subject programs compared with a random, greedy algorithm, and method-total algorithm. Even if the performance of StatPriori showed better performance on both random and coverage-based prioritization techniques, yet this result could be changed when it is implemented on other subject programs and with other coverage-based programs. However, this study could be used as a frame of reference to make wider exploration on the area.

References

1. Mishra, D. B., Panda, N., Mishra, R., & Acharya, A. A. (2019). Total fault exposing potential based test case prioritization using genetic algorithm. *International Journal of Information Technology, 11*(4), 633–637.
2. Software & Systems Engineering Standards Committee I IEEE Computer Society. https://www.computer.org/volunteering/boards-and-committees/standards-activities/committees/s2esc. Accessed 16 Aug 2020.
3. Mishra, D. B., Mishra, R., Acharya, A. A., & Das, K. N. (2019). Test case optimization and prioritization based on multi-objective genetic algorithm. In *Harmony search and nature inspired optimization algorithms* (pp. 371–381). Springer/H. Pham, Software Reliability/Wiley.
4. Mukherjee, R., & Patnaik, K. S. (2019). Prioritizing JUnit test cases without coverage information: An optimization heuristics based approach. *IEEE Access, 7*, 78092–78107. https://doi.org/10.1109/ACCESS.2019.2922387
5. Yoo, M. H. S. (2010). Regression testing minimization, selection and prioritization: A survey. *Software Testing, Verification and Reliability, 7*, 1–30. https://doi.org/10.1002/000
6. Dan, H., Lu, Z., & Hong, M. (2016). Test-case prioritization: Achievements and challenges. *Frontiers of Computer Science, 10*(5), 769–777. https://doi.org/10.1007/s11704-016-6112-3
7. Di Nucci, D., Panichella, A., Zaidman, A., & De Lucia, A. (2020). A test case prioritization genetic algorithm guided by the hypervolume indicator. *IEEE Transactions on Software Engineering, 46*(6), 674–696. https://doi.org/10.1109/TSE.2018.2868082
8. Khatibsyarbini, M., Isa, M. A., Jawawi, D. N. A., Hamed, H. N. A., & Mohamed Suffian, M. D. (2019). Test case prioritization using firefly algorithm for software testing. *IEEE Access, 7*, 132360–132373. https://doi.org/10.1109/ACCESS.2019.2940620
9. Farooq, F., & Nadeem, A. (2017). A fault based approach to test case prioritization. In *Proceedings – 2017 international conference on frontiers of information technology. FIT 2017* (pp. 52–57). https://doi.org/10.1109/FIT.2017.00017
10. Lei, X., Huaikou, M., Weiwei, Z., & Shaojun, C. (2017). *An empirical study on clustering approach combining fault prediction for test case prioritization* (pp. 815–820). IEEE Computer Society.
11. Fu, W., Yu, H., Fan, G., Ji, X., & Pei, X. (2018). A regression test case prioritization algorithm based on program changes and method invocation relationship. In *Proceedings – Asia-Pacific software engineering conference. APSEC* (Vol. 2017, pp. 169–178). https://doi.org/10.1109/APSEC.2017.23
12. Ramya, P., Sindhura, V., & Vidya Sagar, P. (2018). Clustering based prioritization of test cases. In *Proceedings of international conference on inventive communication and computational technologies. ICICCT 2018* (pp. 1181–1185). https://doi.org/10.1109/ICICCT.2018.8473253

13. Pradhan, D., Wang, S., Ali, S., Yue, T., & Liaaen, M. (2018). REMAP: Using rule mining and multi-objective search for dynamic test case prioritization. In *Proceedings – 2018 IEEE 11th international conference on software testing, verification and validation. ICST 2018* (pp. 46–57). https://doi.org/10.1109/ICST.2018.00015
14. Y. Ren, B. B. Yin, and B. Wang, "Test case prioritization for GUI regression testing based on centrality measures," Proceedings of international computer software and applications conference, vol. 2, 61402027, pp. 454–459, 2018, doi: https://doi.org/10.1109/COMPSAC.2018.10275.
15. Luo, Q., Moran, K., Zhang, L., & Poshyvanyk, D. (2019). How do static and dynamic test case prioritization techniques perform on modern software systems? An extensive study on GitHub projects. *IEEE Transactions on Software Engineering, 45*(11), 1054–1080. https://doi.org/10.1109/TSE.2018.2822270
16. Dan, H., Lu, Z., Lei, Z., Yanbo, W., Xingxia, W., & Tao, X. (2016). To be optimal or not in test-case prioritization. *IEEE Transactions on Software Engineering, 42*(5), 490–504.
17. Fazlalizadeh, Y., Khalilian, A., Abdollahi Azgomi, M., & Parsa, S. (2009). Prioritizing test cases for resource constraint environments using historical test case performance data. In *Proceedings – 2009 2nd IEEE international conference on computer science and information technology. ICCSIT 2009* (pp. 190–195). https://doi.org/10.1109/ICCSIT.2009.5234968
18. Do, H., Rothermel, G., & Kinneer, A. (2006). Prioritizing JUnit test cases: An empirical assessment and cost-benefits analysis. *Empirical Software Engineering, 11*(1), 33–70. https://doi.org/10.1007/s10664-006-5965-8
19. Zhou, J., & Hao, D. (2017). Impact of static and dynamic coverage on test-case prioritization: An empirical study. In *Proceedings – 10th IEEE international conference on software testing, verification and validation workshops. ICSTW 2017* (pp. 392–394). https://doi.org/10.1109/ICSTW.2017.74
20. Ammar, H. H., Nikzadeh, T., & Dugan, J. B. (1997). A methodology for risk assessment of functional specification of software systems using colored Petri nets. In *Proceedings of the 4th international symposium on software metrics* (p. 108).
21. Munson, J. C., & Khoshgoftaar, T. M. (1996). Software metrics for reliability assessment. In *Handbook of software reliability engineering* (pp. 493–529). McGraw-Hill, Inc.
22. Halstead, M. H. (1977). *Elements of software science*. Elsevier North-Holland, Inc.
23. Henard, C., Papadakis, M., Harman, M., Jia, Y., & Le Traon, Y. (2016). Comparing white-box and black-box test prioritization. In *Proceedings of international conference on software engineering* (pp. 523–534). https://doi.org/10.1145/2884781.2884791

A Literature Review on Software Testing Techniques

Kainat Khan and Sachin Yadav

1 Introduction

The evolution and making of big software products include various tasks that must be synchronized to fulfil desired needs. Software testing is a crucial task in SDLC phases. It's a method of assessing software in order to identify bugs/errors in the program. Software testing is performed to analyze if the software product fulfils the specified quality requirements or not, and there are a series of steps in which it is developed to ensure that the code of the program does what it specifies. It also checks and validates the working of a software program. Software testing is done to check the other quality factors (ability/potential, trustworthiness, righteousness, constancy, usefulness, proficiency, cost-effectiveness, transferability, maintainability, similarity, etc.). For the past years and till now, everyone is working with similar techniques. Testing is an expensive process as it requires skilled testers as well as advanced and latest technology tools. Testing a software, targets attaining specific objectives and principles that need to be chased. Some of the objectives of testing involves the following:

- The better the software performs, the more effectively testing is carried out.
- When there are less changes in software code, there will be less interruption and delays [16].
- Demonstrate: Testing illustrates functions that are under some specific conditions/rules and coveys that the software product is ready to use.
- Detect: It identifies faults and the errors present in it [8].

K. Khan (✉)
Department of Computer Science, Delhi Technological University, Delhi, India

S. Yadav
School of Engineering Technology, Noida International University, Noida, Uttar Pradesh, India

M. Khari et al. (eds.), *Optimization of Automated Software Testing Using Meta-Heuristic Techniques*, EAI/Springer Innovations in Communication and Computing, https://doi.org/10.1007/978-3-031-07297-0_5

- Prevention: It presents data to curb and decrease the errors to improve system performance.
- Improvement in quality: With the support of efficient testing, the software quality upgrades.

2 Review Methodology

This section involves the study of methods used and presented in the paper. The whole research work presented in this work is partitioned into various phases, that is, the initial stage where planning is done, and based on that many research questions (RQ) were framed, and then in the next phase, research works were shortlisted. In the final phase, that is, third stage, all the research questions were answered (Fig. 1).

2.1 Planning

The test suite contains a huge amount of test cases, and executing them is an annoying task. Many researches have been done to minimize this annoying task. The test suite is minimized by removing the redundant test cases and removing those cases which are no longer needed, or the functionality has been removed in the updated version of the software.

2.1.1 Searching Process

The search process starts by going through a lot of different online sites, search engines, and journals. Many research works published by scholars were explored with the help of IEEE Xplore, Springer, Google Scholar, and Scopus. Relevant

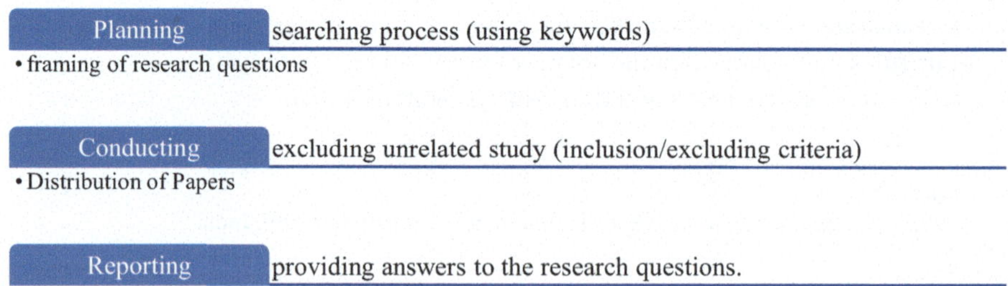

Fig. 1 Review process

keywords like "software testing techniques" and "software testing tools" were applied to fetch similar works, and in this study, the search was limited to the time period between 2007 and 2020. The search process is done by framing and using keywords that are given below:

```
    (SoftwareTesting      OR      SoftwareTestingTechniques      OR
(TechniquesUsedInSoftwareTesting) AND (SoftwareTestingTools OR
 TestingTools)
```

After the completion of the search processes, the next step is to frame the research questions.

2.1.2 Formulation of Research Questions

Many RQs were uplifted while determining and evaluating the preselected studies. All the RQs were appropriately framed to preserve the flow of information in the study and to circumvent discrepancy while studying the broad information gained from the shortlisted papers. Table 1 gives the series of RQs [2, 4, 15].

2.2 Conducting Phase

This portion involves the tasks conducted in the performing phase of literature review. This section also provides the criteria applied while filtering out the studies based on their relevance. Additionally, a research questionnaire was prepared to represent the study conducted. Graphs and pie charts were made to represent the selected studies.

Table 1 Research questions

S. no	Research questions
1.	What is the need and benefit of software testing [10, 16, 17]?
2.	What are the principles of software testing [11, 22]?
3.	What are the phases involved in software testing life cycle [15]?
4.	What are the various types of software testing [15, 24]?
5.	What are various software testing techniques [1, 29]?
6.	What types of bugs are present in software testing [9, 15]?
7.	What are the latest software testing tools [1, 22]?
8.	What are the different test case design techniques [28]?

2.2.1 Excluding Unrelated Study

All the research work is maintained in a sequential manner, and ambiguous and unrelated studies were neglected. As this work is related to software testing techniques, there were an acceptance and excluding criteria to shortlist papers based on software testing techniques.

Acceptance criteria are as follows:

1. Studies that relate to software testing methodologies
2. Studies that mentioned different software testing tools
3. Studies that relate to STLC (software testing life cycle)
4. Studies that cover the major portion of software testing types and the test case design techniques

Excluding criteria are as follows:

1. Studies that didn't target software testing techniques and types
2. Studies that belong to the years other than the time period of 2007–2020
3. Studies that do not related to the framed research questions (as mentioned in table above)

2.2.2 Distribution of Papers

The research studies were partitioned into three parts based on the source of their publication. The sources taken into consideration were journals, conference, and some other work like workshops, chapters, symposium, etc. Initially 80 research papers were downloaded, and from those 80 papers, 49 studies were shortlisted. The division of work is shown by using pie chart. The pie chart is based upon the overall research paper studies (not on the shortlisted ones) (Fig. 2).

- As Per Year of Publication

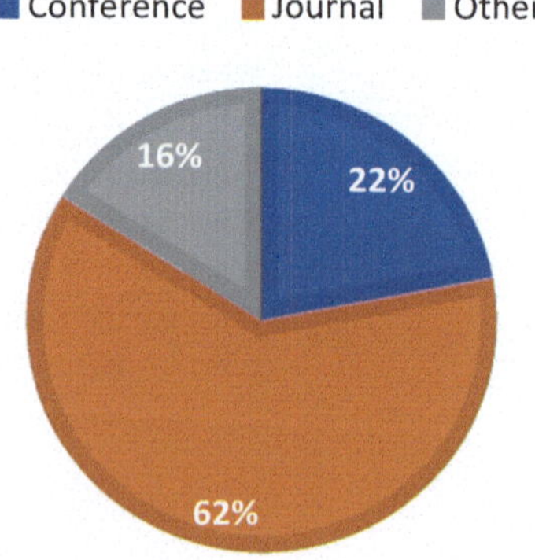

Fig. 2 Distribution of papers

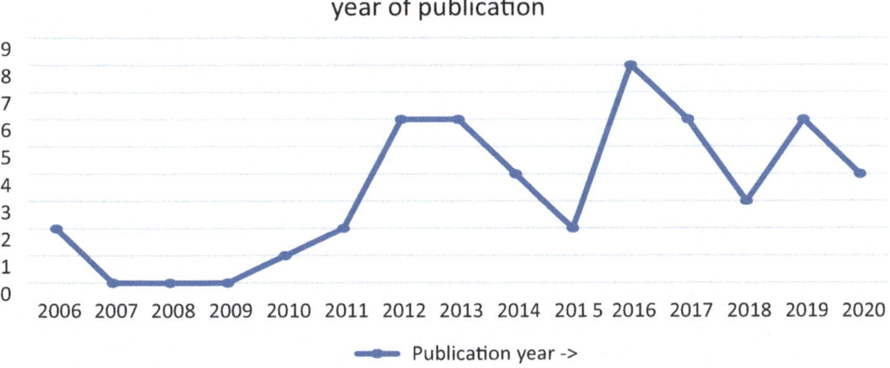

Fig. 3 Arrangement of papers on the basis of their publication year

Multiple studies have been done since past years in the field of "software testing techniques." The selected research works lie in the years 2006–2020 (Fig. 3).

2.3 *Reporting of Research Questions*

RQ1: What are the needs and benefits of software testing?

Software generation includes developing a software with set specification. Software testing procedure is essential to verify as well as to validate the resulted software (to look if it meets all requirements specified, otherwise we will lose our customer). To make sure that we give all of our customers an appropriate software solution, we opt for testing. Testing guarantees that the end product exactly matches the set of requirements and what we actually wanted to make [19, 25].

The demand for conscientious testing and its related work during the SDLC phase appears due to the mentioned reasons [18]:

- To detect defects.
- To decrease imperfection in the system.
- To hike the comprehensive standard/quality of the system.

These points below tell us the importance of testing for an authentic and useful software product [7].

- To acquire user insight: Software testing ensures that the software made is easy to use (user-friendly). Testers who are specialized in testing software know the requirements of users, and if the software cannot satisfy the user's requirement, then it is useless investment [20].
- To test software ability: A lot of different devices, OS, and browsers are present, so it is essential to check whether the software is compatible with others or not, and this step gives a smooth and good user experience.

- To detect errors: Testing a software helps to get rid of errors before the product delivery because the chances of having any kind of error are always present.
- To hike the software development process: Spending a lot of time on the development process is not a good practice. Software should be timely deliverable [4].
- To keep away software from risk: Software must be free from loopholes; otherwise it will lead to loss of data.
- To enhance business: When the software is of good quality, it helps the brand to gain good reputation as well as profit.
- Cost-effective: It is also a crucial benefit of testing software. Additional charges may give the user a bad experience. Testing before delivering helps the organization to save a lot of bucks because when we identify bugs in the early stage, the cost of fixing it gets reduced.
- Security: The most important benefit because everyone looks for software that are trustworthy [12].

RQ2: What are the principles of software testing?

Principle is a process in action which must be tracked. Testing principles are mentioned below:

1. Test all your program/software with the intent of making it fail or deteriorate: Testing is the method of implementing a program/software in order to find problems in it, and we must do testing in such a way that it exposes all the failures present in that software to make the process of testing more fruitful and valuable[13].
2. Testing must be done at an early stage: Testing that has been done in the early phase helps in improving the quality and fixes enormous errors.
3. Testing dependent on the context: Testing has to be proper to serve the desired purpose. Various testing must be done at different time intervals [29].
4. Plan clear test cases: Test case should be framed in such a format that is quantifiable, which will lead to unambiguous testing results [3].
5. Check for valid and invalid input conditions: Along with the present valid inputs, the invalid/unexpected conditions should also be checked.
6. Testing should be carried out by various people at various stages: Various motives are tackled at various stages of testing, so it is necessary to perform testing at regular intervals.
7. One has to stop the testing process at some point. We can put an end to testing when there is a curb [23].

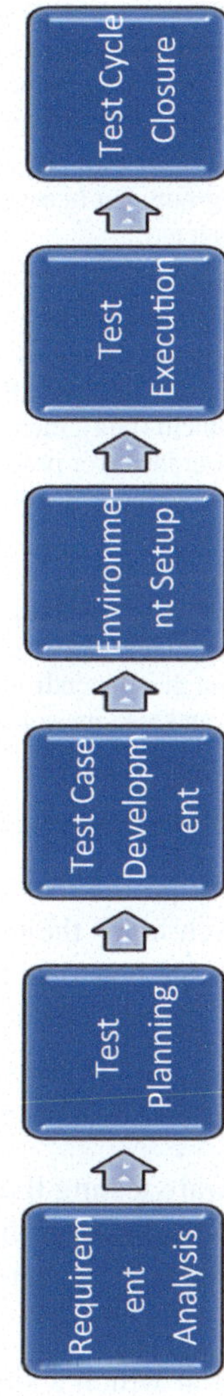

Fig. 4 Life cycle of software testing

RQ3: What are the phases involved in software testing life cycle? (Fig. 4)

1. *Requirement analysis*	This is the initial step in the STLC process, and here, the quality assurance group/team acknowledges the fundamental needs or requirements of the customer and determines the requirements that are testable, and in the future, if any ambiguity arises, the QA team again collaborate and rectify the problem. This part creates a framework for the test plan.
2. *Test planning*	It is an essential and compulsory step as the test strategy is explained in this phase because without this testing cannot be done. Test planning is more inclined toward the functionality testing.
3. *Test case development*	Details of test cases are mentioned and created at this point. Accurate test cases are made by QA group, but in several cases automatically generated test cases are also considered.
4. *Environment setup*	Setting up of required software as well as hardware for testing team so that the created test cases could implement.
5. *Test execution*	It includes implementation of test cases depending upon the previously made test plan (as it behaves as input). If the program clears this phase without having any problem (bug), then it is said to be cleared; otherwise, we will get to know about the error present in it. We process the code and match the desired output with our result. The ultimate deliverable of this phase is error or bug.
Test closure	It is the final step involved in STLC. An analytical review is done in this phase. The results of implemented test plan are validated, and further, the decision is taken. The team that tests collaborate in order to analyze criteria for cycle completion and that depends upon factors like quality, objective of the business, coverage, and also the software.

RQ4: What are the various types of software testing?

Testing is a fundamental phase of a productive and booming software project. Software testing is of many types which helps the tester to choose the most appropriate testing technique for the project that has to be built. The testing done by quality assurance team/software tester is of two types: functional and nonfunctional (Fig. 5).

- *Functional Testing*

Functional testing is a part of software testing that validates whole functionality of the software built in accordance with the specified requirements. This portion targets on automated along with manual testing. Various types of functional testing are as follows:

1. *Unit Testing*: It is that stage of test at which a small single unit of a software is tested. The main aim of unit testing is validating that every individual component works and behaves as designed. A unit/component has mostly one or sometimes more than one input and generally one output [32].

   Advantages

   (a) Debugging of software is not complicated.

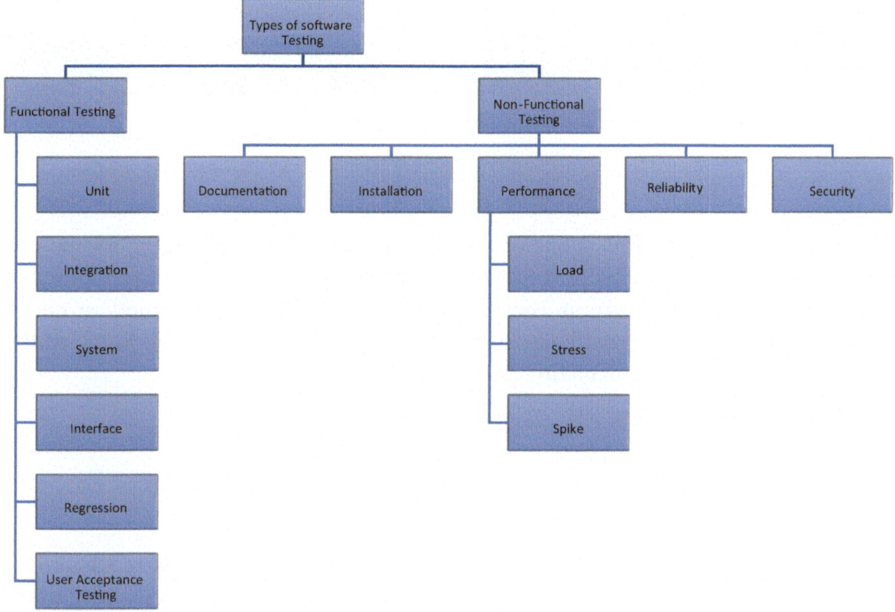

Fig. 5 Types of software testing

(b) Reusable codes are made. Advanced codes are required for making it possible.
(c) Speedy development is done as well as less efforts are needed.
(d) Low cost is needed to fix defects.

2. *Integration Testing*: At this stage of testing, all the previously made single units/ components are put together as they are tested as a whole combined group. The main motive of doing integration testing is to identify and analyze faults present between the units that are integrated.

   Various methods are used for doing such type of testing:

   (a) Top-down: When the uppermost (top level) units are tested and then the lower-level units are tested, it is said to top-down approach.
   (b) Bottom-up: It is just the opposite of top-down testing as in here lower-level units are tested before the top-level units.
   (c) Sandwich: It's a fusion of both upper-level and bottom-up techniques.

3. *System Testing*: At this point of software testing, the complete software is tested, and this phase helps analyzing the software with the given requirements.
4. *Interface Testing*: Software consists of various component, and those could be anything like DB or a server. The network that connects and provides a facility of communication among different units is called as interface. It checks whether the communication is done properly or not (Fig. 6).

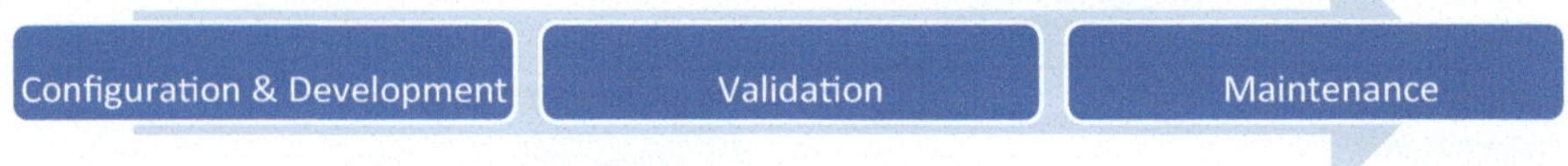

Fig. 6 Stages of interface testing

5. *Regression Testing*: This is one of the most essential levels of testing for the end product and also very helpful for the programmers to look for the reliability of the object with the current needs. Regression testing is performed to check whether the change in the code influences the functionality or not. Some of the techniques of regression testing are as follows:

   (a) Retest all test cases
   (b) Regression test selection
   (c) Prioritizing test cases on the basis of its influence and criticality
   (d) Hybrid

6. *User Acceptance Testing*: Here we test the software to check for its acceptability. The purpose is to identify whether the software is acceptable for delivering to the user or not. Alpha testing (when software is assessed by tester) and beta testing (to check the software in real-world environment) are some types of user acceptance testing.

- *Nonfunctional Testing*

While developing and testing a software, there are some nonfunctional requirements also present, and they include performance, quality, etc.

1. *Documentation Testing*: It is done to approximate the required testing efforts as well as test coverage. The documentation consists of a test plan, test case, and also gathering of needs/requirements [27].

   (a) Instructions: Some predefined rules are given for the associated work, that is, having test scenarios.
   (b) Examples: Procedural examples are given for ease.
   (c) Messages.
   (d) Samples.

2. *Installation Testing*: It is a quality assurance testing that deals with what the user will require to download and how the new software will successfully set up.
3. *Performance Testing*: It is done to verify whether the application is working as expected or not under the given workload.

   (a) Load test: To analyze system behavior in the case of increased work
   (b) Stress test: To analyze system behavior in extreme conditions that are not expected

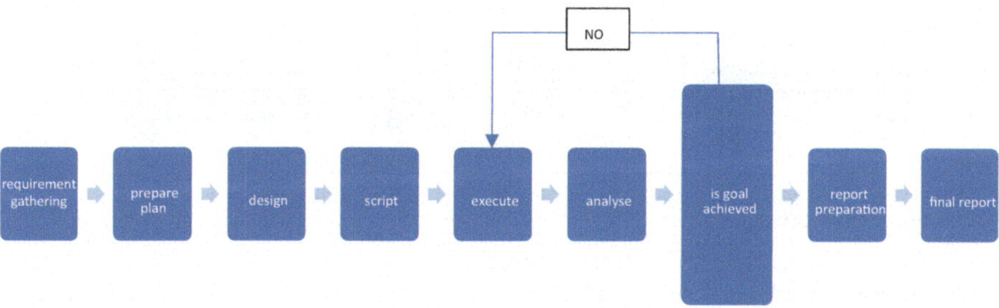

Fig. 7 Steps in performance testing

Fig. 8 Black box testing

(c) Endurance test: To analyze system behavior when notable work is given regularly
(d) Spike test: To analyze system behavior when workload is promptly increased (Fig. 7).

4. *Reliability Testing*: It guarantees that the software is free from faults and is trustworthy for the purpose it made. Here types of testing such as feature test, regression test, load test, etc. are performed.
5. *Security Testing*: In this, testing makes sure that resulted software is free from any kind of loopholes. At this stage, we look for all the possible defects that the software contains because it might lead to a huge loss of data.

RQ5: What are various software testing techniques?

The main software testing techniques are as follows:

(1) Black Box Testing

This specific testing is performed to check/test the functioning of a program. The alternative name of black box testing is behavioral or specification testing. In this, we have a set of values, that is, the input values and the desired output values.

Input is given, and if its result is equivalent to the output we wanted, then the working is said to be "ok"; otherwise there is some problem in the code. While performing black box testing, the only thing that is known to the person who tests is input values set and their output, but the internal functionality remains unknown (Fig. 8).

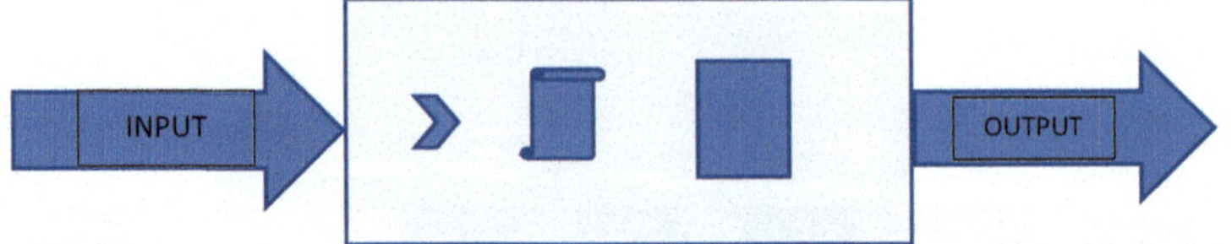

Fig. 9 White box testing

The main aim of performing black box testing is to look for absent/incomplete values, errors in performance, and also the errors that occur when the external DB is accessed [9].

Advantages	Disadvantages
Tests are not biased	When requirements are not specified clearly, designing of test cases gets difficult
As soon as specifications are completed, the test cases are designed	For complex code structure black box testing cannot be performed
Accessing the code segment is not needed	Due to the limited access and knowledge, this becomes inefficient

(2) White Box Testing (Fig. 9)

It is done to check/test software and the way it is implemented so that the efficiency and its structure could improve. The alternate names of white box testing are structural/clear-box/transparent box/open box testing. Here, the tester knows whole design/ internal structure of the program. The functioning of white box testing is opposite of the black box. Because of the known components in white box testing, it becomes simple and easy to find errors even when the software specifications are incomplete. The motive of performing this testing is to make sure that all the test cases include each and every path by going through the program. This further assures that internal units are made as per the specified design [21].

Advantages	Disadvantages
When testing is performed, it accumulates bigger portion of the program code	Because it covers a major portion of coding, it might not be suitable for accessing functionality
It does not cover any error that is typographical	As it is design based, there is a chance that it does not focus on other problems occurring in the system
It also identifies all the errors made while designing	Whenever there is a modification in implantation, all the test cases have to be modified

RQ6: What types of bugs are present in software testing?

Bugs in software testing are categorized into three parts, and those are defects occurred by nature, due to priority and due to severity [26].

- Software Bugs by Nature:

  1. Functional bugs: These are the problems analyzed when the software behavior is not adaptable with the specified needs/requirements. These are identified while performing functional testing [13].
  2. Performance bugs: Bugs that are confined to the speed of the software, its feedback/response time, its stability, and consumption of resource. These are identified while doing performance testing. For example, if response time > specified time (as given in requirements), it is said to performance bug.
  3. Usability bugs: This type of bugs makes the software inappropriate and problematic to use which leads to bad user experience.
  4. Compatibility bugs: A software with compatibility problems is inconsistent in terms of performance on specific device, OS, hardware, and browsers. To tackle and to find out such problems, compatibility testing is done [27].
  5. Security bugs: This relates to the weakness of a software that allows attackers to break security of software.

- Software Bugs due to Severity:

  1. Critical problems/defects: They obstruct the whole functionality of a system which forces testing to stop without the bug being resolved.
  2. High severity bugs: They affect the main functioning of software application, and they work in a way that is opposite as stated in requirement.
  3. Medium severity bugs: When a minor portion of the function does not work properly.
  4. Low severity bugs: It is basically linked to software application's user interface.

- Software Bugs by Priority:

  1. Urgent defects: Such defects must be resolved within a specified time which is usually 24 hours after it has been reported. Problems that are crucial lie in this group.
  2. High priority bugs
  3. Medium priority bugs
  4. Low priority bugs

RQ7: What are the latest testing tools?

1. Selenium: It acts as a base for a lot of tools used in software testing.

   It provides a structure/framework to carry out testing among different browsers/platforms (Linux, windows, etc.).
   It has some extra features in it like recording and playback.
   Testers can write test cases in different programming language as it is supported by selenium [30].

2. TestComplete: It is a platform for performing functional testing. It provides results for automate testing.

    It also offers features of recording and playback like selenium does. It has some additional features like GUI testing.

3. TestingWhiz: It is an automation tool.

    It authorizes to automate user interface and functions to examine the functionality of applications.
    It delivers organized interface to provide effective user experience.
    It permits to check the DB of an application by various methods [31].

4. Ranorex: It provides a lot of different automation tools for testing all kinds of mobile and desktop applications.

    It offers features like bug detection.
    In Ranorex test codes are reusable.
    It has features like GUI recognition and image comparison.
    It can quickly make, implement, and deliver a reliable automated GUI (with/without coding).
    Test creations can be made codeless.
    Good and efficient customer support.

5. Sahi: It is used to automate application on web.

    It is an open-source tool.
    It supports multi-browser testing.
    Like other tools, it also has features of recording and playback.

RQ8: What are the different test case design techniques?

Designing a test case means the way in which test cases are set up. The main motive of designing test cases is to check the functioning and its features.

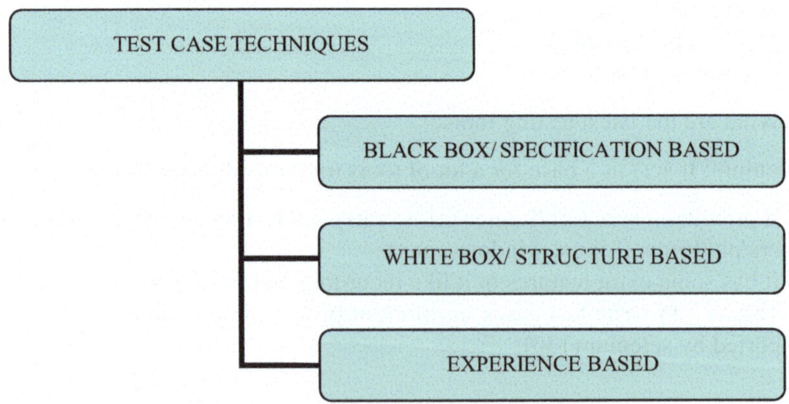

Fig. 10 Test case techniques

They are classified into three parts (Fig. 10):

- *Specification Based*

1. Boundary Value Analysis (BVA): BVA helps in exposing errors present at the edges of input. The input values are partitioned into maximum and minimum values. It finds out errors that obstructs the functionality of a program.
2. Equivalence Partitioning: In this, the test data (input) is divided into various classes with equal number of information/data [27].

   In the next step, for each partition, the test cases are designed. It permits to find valid/invalid classes.
3. Decision Table: The alternative name of decision table testing is cause-effect table testing. Based on the decision table, test cases are made which were generated with the help of various combinations of data (input and output data).
4. State Transition: When there is a change in the conditions of input, the respective AUT also changes. This helps to check the performance of AUT.
5. Use Case Testing: It is a specification of a specific use of program by the customer. Various test cases are designed to tackle various scenarios.

- *Structure-Based Testing*

  1. Statement Testing and Coverage: Each and every statement in the software program code are implemented at least once. As per the specified requirements, the percentage of statement is computed [24].
  2. Decision Testing Coverage: The alternative name of decision testing coverage is branch coverage. In this, from every decision point all possible branches are executed at least once. This assists in validation of every branch in code.
  3. All Path Testing: This testing helps in finding and analyzing each and every fault in the code.

- *Experience-Based Testing*

  This technique depends upon the experience of the tester.

1. Guessing the Error: The tester finds faults and errors depending upon their past experience, data available, and knowledge.

3 Conclusion

Testing is one of the most crucial sections of SDLC, because the end delivery of software product is dependent upon it. Testing a software is not a simple process because it needs a lot of time and effort as it is an exhaustive process. It requires advanced and latest technologies. With the help of the latest software testing tools and technologies, testing is performed more appropriately [6].

- Testing a software is the most fundamental task of software engineering.

- Testing is performed to make sure and to raise the likelihood of profit rate of software product.
- Testing is a continuous and endless process.
- Testing only identifies the existence of faults/errors, not its exclusion.
- It is a task that implements software code with the motive of analyzing and exploring errors in the code.
- This work outlines the techniques used in software testing and various strategies of it.

References

1. Anitha, A. (2013). *A brief overview of software testing techniques and metrics*. [Online]. Available: www.ijarcce.com
2. Anwar, N., & Kar, S. (2019). *Review paper on various software testing techniques & strategies*.
3. Atifi, M., Mamouni, A., & Marzak, A. (2017). A comparative study of software testing techniques. *Lecture Notes in Computer Science (including subseries Lecture Notes in Artificial Intelligence and Lecture Notes in Bioinformatics), 10299*, 373–390., LNCS. https://doi.org/10.1007/978-3-319-59647-1_27
4. Babbar, H. (2017). *Software testing: Techniques and test cases*. [Online]. Available: www.ijrcar.com
5. Baresi, L., & Pezzè, M. (2006). An introduction to software testing. *Electronic Notes in Theoretical Computer Science, 148*(1 Special Issues), 89–111. https://doi.org/10.1016/j.entcs.2005.12.014
6. Chaudhary, S. (2017). Latest software testing tools and techniques: A review. *International Journal of Advanced Research in Computer Science and Software Engineering, 7*(5), 538–540. https://doi.org/10.23956/ijarcsse/sv7i5/0138
7. Divyani, M., Taley, S., & Pathak, B. (n.d.). *Comprehensive study of software testing techniques and strategies: A review*. [Online]. Available: www.ijert.org
8. Garousi, V., & Mäntylä, M. V. (2016). A systematic literature review of literature reviews in software testing. *Information and Software Technology. Elsevier, 80*, 195–216. https://doi.org/10.1016/j.infsof.2016.09.002
9. Iqbal Malik, K., & Hassan, S. (2013). *Software testing methodologies for finding errors software architecture refactoring tools and techniques: a comparative study view project*. [Online]. Available: www.theinternationaljournal.org
10. Jamil, M. A., Arif, M., Abubakar, N. S. A., & Ahmad, A. (2017). Software testing techniques: A literature review. In *Proceedings – 6th international conference on information and communication technology for the Muslim world, ICT4M 2016* (pp. 177–182). https://doi.org/10.1109/ICT4M.2016.40
11. Jammalamadaka, K., & Parveen, N. (2019). Holistic research of software testing and challenges. *International Journal of Innovative Technology and Exploring Engineering, 8*(6 Special Issue 4), 1506–1521. https://doi.org/10.35940/ijitee.F1307.0486S419
12. Jat, S., & Sharma, P. (2017). Analysis of different software testing techniques. *International Journal of Scientific Research in Research Paper. Computer Science and Engineering, 5*(2), 77–80. [Online]. Available: www.isroset.org
13. Jovanović, I. (n.d.) *Software testing methods and techniques*.
14. Kaur Chauhan Á, R., & Singh Ḃ Á, I. (2004). *Latest research and development on software testing techniques and tools*. [Online]. Available: http://inpressco.com/category/ijcet
15. Kaur, M., & Singh, R. (2014). *A review of software testing techniques*. [Online]. Available: http://www.irphouse.com.

16. Kumar, R. (2016). *Software testing techniques and strategies*. [Online]. Available: http://www.ijeast.com
17. Malviya, A. (n.d.) *Software testing: Concepts and issues*. [Online]. Available: https://ssrn.com/abstract=3351067
18. Naik, K., & Tripathy, P. (2011). *Software testing and quality assurance: Theory and practice*.
19. Nidhra, S. (2012). Black Box and White Box testing techniques – A literature review. *International Journal of Embedded Systems and Applications, 2*(2), 29–50. https://doi.org/10.5121/ijesa.2012.2204
20. Okezie, F., Odun-Ayo, I., & Bogle, S. (2019). A critical analysis of software testing tools. *Journal of Physics: Conference Series, 1378*(4). https://doi.org/10.1088/17426596/1378/4/042030
21. Orso, A., & Rothermel, G. (2014). Software testing: A research travelogue (2000–2014). In *Future of software engineering, FOSE 2014 – Proceedings* (pp. 117–132). https://doi.org/10.1145/2593882.2593885
22. Pardeshi, S. N. (2013). Study of testing strategies and availabletools. *International Journal of Scientific and Research Publications, 3*(3) [Online]. Available: www.ijsrp.org
23. Pezzè, M., & Young, M. (2008). *Software testing and analysis: Process, principles, and techniques*. Wiley.
24. Rosero, R. H., Gómez, O. S., & Rodríguez, G. (2016). 15 years of software regression testing techniques – A survey. *International Journal of Software Engineering and Knowledge Engineering. World Scientific Publishing, 26*(5), 675–689. https://doi.org/10.1142/S0218194016300013
25. Roshan, R., & Sharma, C. M. (2012). *Review of search based techniques in software testing*. Rabins Porwal ITS.
26. Sawant, A. A., Bari, P. H., & Chawan, P. M. (2015). *Software testing techniques and strategies* (Vol. 2, pp. 980–986) [Online]. Available: www.ijera.com
27. M. A. Sethi (n.d.), "A review paper on levels, types & techniques in software testing," International Journal of Advanced Research in Computer Science 8, 7, https://doi.org/10.26483/ijarcs.v8i7.4236.
28. Sharma, C., Sabharwal, S., Sibal, R., Hind, A., & Marg, F. (2013). *A survey on software testing techniques using genetic algorithm*. [Online]. Available: www.IJCSI.org
29. Singh Ghuman, S. (2014). *International Journal of Computer Science and Mobile Computing Software Testing Techniques*. [Online]. Available: www.ijcsmc.com
30. Singhal, A., Bansal, A., & Kumar, A. (2013). A critical review of various testing techniques in aspectoriented software systems. *ACM SIGSOFT Software Engineering Notes, 38*(4), 1–9. https://doi.org/10.1145/2492248.2492275
31. II. TESTING TYPES. (2013). [Online]. Available: http://www.toolsjournal.com/testinglists/item/404-10-
32. Vos, T. E. J., Marínt, B., Escalona, M. J., & Marchetto, A. (2012). A methodological framework for evaluating software testing techniques and tools. In *Proceedings – International conference on quality software* (pp. 230–239). https://doi.org/10.1109/QSIC.2012.16

A Systematic Literature Review of Predicting Software Reliability Using Machine Learning Techniques

Getachew Mekuria Habtemariam, Sudhir Kumar Mohapatra, Hussien Worku Seid, and Deepti Bala Mishra

1 Introduction

During the development process of software life cycle, it is almost difficult to avoid the injection of fault in software development. So, it is essential to identify and solve the faults even the failure as much as possible before delivering the software to the end user [1]. The objective of software verification and validation is to identify errors and fix as much as possible. As more and more errors are discovered, there is a need to be fixed through testing, and the number of errors found in the software decreases, while on another side, it increases the software reliability. A number of growth models of software reliability have been established in the previous to predict and estimate the reliability of software [2]. These models can also help in determining at what time the testing process can stop and deliver the software to the end user. The objective of this review is to identify and investigate the different previously conducted research paper on software reliability prediction using different machine learning techniques used to predict errors in the life cycle of software development which is studied between 2010 and 2020. Our reviews investigate how different machine learning techniques perform during software development of early prediction of software fault. Our results enable researchers to develop software reliability estimation and prediction via machine learning methods based on best experience and practice across many previous studies. In addition

G. M. Habtemariam · H. W. Seid
Addis Ababa Science and Technology University, Addis Ababa, Ethiopia
e-mail: hussien.seid@aastu.edu.et

S. K. Mohapatra (✉)
Faculty of Emerging Technologies, Sri Sri University, Cuttack, Odisha, India
e-mail: sudhir.mohapatra@srisriuniversity.edu.in

D. B. Mishra
Department of MCA, GITA Autonomous College, Bhubaneswar, India

M. Khari et al. (eds.), *Optimization of Automated Software Testing Using Meta-Heuristic Techniques*, EAI/Springer Innovations in Communication and Computing, https://doi.org/10.1007/978-3-031-07297-0_6

to this, our review output also helps reader to make efficient decisions on software estimation and prediction most suited to their context [3, 4].

The emergence of a huge number of software reliability growth models alone is a manifestation that there is no specific model appropriate for predicting the reliability of all types of software. Therefore, in the future one of the research areas that can be establish a specific model appropriate for all kinds of state of affairs or establishment of a joined up or universal model which can acquire forms appropriate for diverse types of the software.

Another significant research area can be detecting the best-tailored model for a given scenario. Several investigation standards have been proposed for stating the best-tailored model. Several times, the different measurement may propose different models as the best-tailored model for a given scenario. Therefore, the establishment of a procedure to state the best model tailored for a given scenario based on several measurements can be additional area of upcoming research.

Guarantee of software reliability is an essential issue in the establishment of software applications. Software reliability is among the vital component of software quality which affects the quality of the software [2]. In a broad sense, software quality consists of many features, such as software functionality, availability, maintainability, robustness, performance safety, etc. [1, 5]. Software reliability considers into account all kinds of errors and defect that may cause a tough risk in any types of software programs. To evaluate and forecast the software reliability with appropriate accuracy is a difficult task.

The organization of the remaining part of the review is as follows: Sect. 2 shows the procedure as well as the research questions which are handled in this review and followed by the research criteria for choice of primary studies. Section 3 portrayed the boundary and the restriction of the review work that exhibits the responses to the research questions which are stated in this work. Section 4 states the result or outcome. Finally, the conclusion and some future directions are drawn in Sect. 5.

2 Methodology for the Review

The overall activities of organizing, directing, and broadcasting of the systematic review conducted in this review research are carried out by following the standard furnished by Kitchenham [1]. For a simple understanding, the step of systematic literature review is illustrated in Fig. 1. In the initial step, we established the review guideline which enclosed the following line of actions: finding out research questions, plan search schemes, analysis selection procedures, investigate quality assessment, data pulling out operation, and data integration procedure. Following the establishment of the review standard, a series of processes was conducted in the review. In the initial stage, we established the research questions which managed the problems to be answered in the literature review. In the next stage, we portrayed the search scheme which incorporates the pinpointing of search terms and choice of sources to be found in order to identify the key studies.

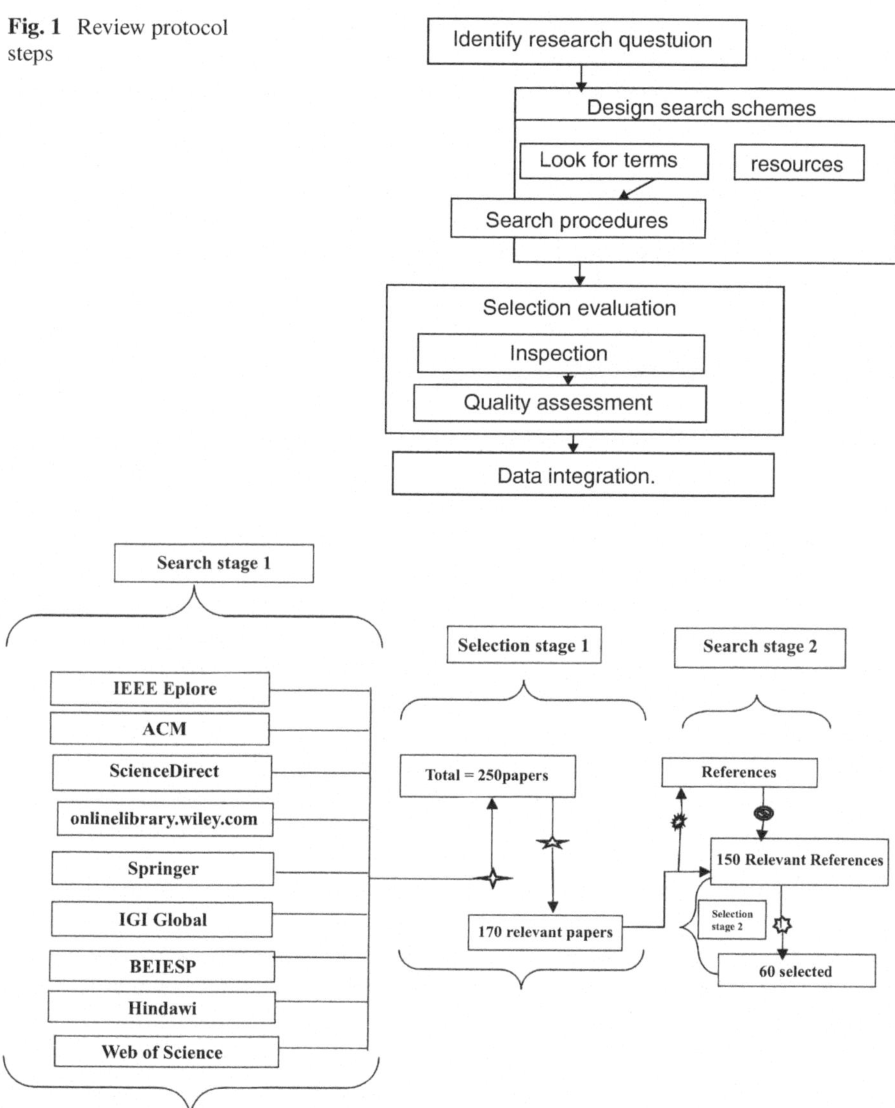

Fig. 1 Review protocol steps

Fig. 2 Search and selection process

Following the next step is the designation of appropriate analysis based on the research questions. This stage also controls the inclusion and exclusion guideline for each primary study. After following the next step, we state quality assessment scheme by creating the quality assessment surveying in order to examine and evaluate the studies. The final step engages the design of data pulling out forms to collect the essential information with the aim of responding to the research questions, and in the final step, we plan procedures for data integration, which are portrayed in Figs. 1 and 2.

Establishment of review guideline is an essential step in a systematic literature review as it depletes the likelihood and threat of research unfairness in the systematic literature review. We developed the review scheme in this task by repeatedly holding dialogues and consultation with our adviser which involve a senior associate and assistant professors. In the subsequent sections, we specify the research questions and the steps followed while developing the systematic literature review.

2.1 Research Questions

The purpose of this systematic literature review is to deliver and evaluate the fact-finding evidence acquired from the studies of software reliability using machine learning approach in the literature. Table 1 portrayed 12 research questions handled in this literature review. From the key studies we found the basic ideas and definition of software reliability used for predicting software reliability (RQ1). We investigated these studies using the machine learning approach for software reliability prediction to respond to RQ2, RQ3, RQ4, RQ5, RQ6, RQ7, and RQ8. In the subsequent research question, we investigated the overall efficiency of the machine learning approach for software reliability prediction (RQ9). This question concentrates on the values of the efficiency measures for software reliability prediction. RQ10 analogizes the efficiency of machine learning approach and analytical methods on the software reliability prediction. The objective of this question is to decide if machine learning approach is ahead of the analytical methods. RQ11 evaluates the cross-validation techniques as well as which machine learning approach outshined the other machine learning approach in order to decide which machine learning approach is persuasively ahead of the other machine learning approach. The last question (RQ12) decides the effectiveness and shortcoming to bring software professional direction concerning the choice of the appropriate machine learning categories.

2.2 Search Scheme and Selection Analysis

We established a complicated look for keywords by integrating different terms and synonyms applying Boolean statement "OR" and coupling with the vital look for terms applying "AND." The subsequent common look for terms was applied for distinguishing of main studies:

Software AND (Reliability prediction OR Fault prediction OR defect prediction OR error prediction OR failure prediction) AND (machine learning OR soft computing).

Table 1 Research question

RQ#	Identified RQ (research questions)	Rationalization
RQ1	What are the basic ideas and definition of software reliability?	Assess the different basic ideas and definitions of software reliability discuss by different author.
RQ2	What type of experimental verification used for predicting software reliability?	Explore the different experimental verification which is discussed by different author.
RQ3	Which type of metrics is frequently used for predicting software reliability?	Identify which metrics are commonly used for software reliability prediction.
RQ4	What software reliability data sets are used for build software reliability model?	Assess and evaluate data sets to be appropriate for users to build software reliability prediction model.
RQ5	What are the assessment criteria available to measure the efficiency of software reliability prediction?	Review the evaluation criteria frequently used in each software reliability prediction.
RQ6	Which prediction mode is the best achieving used for software reliability prediction?	Explore the best achieving prediction models in each investigation of the previously conducted research.
RQ7	Which collaborative prediction models are used to predict software reliability?	Examine the collaborative prediction models frequently used for in software reliability.
RQ8	What type of efficiency measurement is used for software reliability prediction?	Explore the type of efficiency measurement used for software reliability prediction.
RQ9	What is the overall effectiveness of the machine language for software reliability prediction?	Investigating the overall effectiveness of the machine learning approach for software reliability prediction.
RQ10	Which technique outshines performance of the machine learning and statiscal methods on software reliability prediction?	Explore the performance of the machine learning technique and stastical techniques on software reliability prediction.
RQ11.	What method of cross-validation is used to assess the efficiency of software reliability prediction models?	Examine the various validation methods applied on software reliability prediction models.
RQ12	Which category of machine learning technique is the best suitable for software reliability prediction?	Explore the different types of machine learning techniques best suitable for software reliability prediction.

Later distinguishing the search terms necessary and significant digital repository were selected. The choice was not limited by the accessibility of digital portals at the home academies. The subsequent nine electronic databases were used for searching the main studies

1. IEEE Xplore.
2. ACM.
3. Science Direct.
4. Wiley Online Library.
5. Springer.
6. IGI Global
7. Blue eye intelligent engineering and science publication.

8. Hindawi.
9. Web of Science.

The search was made from the above nine digital repositories by applying selected search query. We restricted the search from 2010 to 2020 as the systematic literature review of predicting software reliability using machine learning approach was conducted in 2009. After determining which electronic databases to search, an initial search to identify the candidate primary studies was performed. Next to conducting an initial search, the significant studies were decided by acquiring full text papers succeeding the inclusion and exclusion guideline portrayed in the following section. We also incorporate those studies that were found invaluable from the bibliography section of the important studies.

We incorporated the experimental studies of software reliability prediction using the machine learning techniques in the systematic literature review. We also state 250 main studies for inclusion in the systematic literature review. The candidate primary studies were chosen following the inclusion and exclusion criteria which are given below:

Inclusion Criteria

(i) Experimental studies of software reliability prediction using the machine learning techniques.
(ii) Experimental studies coupling the machine language and non-machine language techniques for software reliability prediction.
(iii) Empirical studies comparing the ML and statistical techniques for software reliability prediction.
(iv) For duplicate publications of the same study, only the latest version will be included.
(v) Using hybrid model that employs more than or a least two machine language techniques.

Exclusion Criteria

(i) Evaluate without experimental analysis of software reliability prediction of using of the ML techniques.
(ii) Analysis developed on dependent variables other than fault susceptibility.
(iii) Evaluate applying the machine language techniques in situation other than software reliability prediction.
(iv) Identical investigation, which is the investigation done by the same author in conference and extended version in journal. But if the output were dissimilar in both investigation, they were included.
(v) Software reliability estimation evaluations.

The above guideline of inclusion and exclusion was evaluated by two advisers individualistically, and they arrived a joint decision after detailed discussion. In case of any confusion, full text of the evaluation was inspected, and the last determination concerning the inclusion/exclusion was established. Besides, the quality of the investigation was decided by their significance to the research questions. The identical investigation with the same output by an individual author was removed. By using the above selection guideline, 250 articles were chosen as choice. Lastly, the quality evaluation guideline given in the subsequent section was used to acquire the last investigation.

2.3 *Quality Evaluation Criteria*

The quality evaluation guidelines are executed to assess each investigation developed on the stated research question in the literature review. The quality evaluation guideline follows the stated quality specification as identified by [6]. The primary motto of the quality evaluation guideline is to select investigation and evaluate the investigation which is used to answer our research questions and to assist more detailed studies of inclusion and exclusion guideline. We formed a quality survey for evaluating the importance and effectiveness of the main studies. The quality evaluation is used for weighing the investigation. The quality evaluation queries are stated below.

1. What does the objective of the research clearly state?
2. Is the main research problem defined?
3. Are software reliability and machine learning defined?
4. Is the type of software reliability measurement defined?
5. What does the research methodology clearly specify and is it repeatable?
6. Is the source of the datasets specified?
7. Is a suitable tool for the extraction of the datasets clearly mention?
8. Is the size of the dataset suitable?
9. Is the data collection procedure clear and unambiguous?
10. Does the study ascertain the type of programing language used in the systems being analyzed?
11. Are the independent and dependent variables clearly identified?
12. Is the dataset published publicly?
13. Are appropriate evaluation measures used?
14. Are suitable cross validation techniques used?
15. Are the prediction techniques justified?
16. Are the prediction models and performance of the models properly identified?
17. Are the results and findings clearly discussed?
18. Are the research limitations or challenges properly identified?
19. Does it contribute or add value to the existing literature?

The grading process of the quality evaluate questions is developed based on 1 for yes, 0.5 for partly, and 0 for No. The grade rank of the study groups is as follows: $14 \leq$ score ≤ 15 for excellent, $10 \leq$ score ≤ 13 for good, $5 \leq$ score ≤ 8 fair, and

$0 \leq$ score ≤ 5 for fail [7, 8]. From using the above quality evaluate criteria, 250 papers fail in our quality evaluation. At the end, 60 main paper were selected to conduct for our systematic literature review.

2.4 Data Extraction

The objective of using data pull out is to decide which research question was answered by which main study. During data extraction the researcher performs to pull out extract data from each selected main study with the purpose of collecting data which is used to respond research question. Table 2 classify the categories with regard to our research question necessities.

2.5 Data Synthesis

The aim of data synthesis is to collect and syndicate data from the selected main investigation in order create and reply to respond the research questions. Gathering a number of primary investigations which identify identical and equivalent ideas and output supports in supplying research evidence for acquiring conclusive response to the research questions. This review examined and assessed each of the quantitative data and qualitative data which forms a body of fact from the selected investigations that handle the problems corresponding to our research question [7–9]. Quantitative data includes values of various performance metrics and qualitative data which incorporate the effectiveness and limitation of the machine learning procedures and categorization of various machine learning procedures. With the aim of responding to our research questions, we applied visualization approaches.

Table 2 Data pull out category corresponding with research query

Category	Research question
Over all ideas and definition of software reliability	RQ1
Experimental verification of software reliability using machine learning technique	RQ2
Software reliability metrics	RQ3, RQ4, RQ5
Software reliability data set and size	RQ6, RQ7
Performance evaluation of software reliability prediction model and ensemble model	RQ8, RQ9,
Overall performance machine	RQ10, RQ11, RQ12.

3 Discussion on Some Selected Article

Jaiswal et al. [2] stated and defined software reliability, that is, "The capability of the computer program to achieve its end user requirement operation under identified situation for a quantified period of time." As human need increaser to use software, parallel with the fast progression and rising sophistication of the software. So, developing quality software is tough to acquire.

The researcher applied machine learning techniques for predicting and forecasting software reliability. But as need, complexity, and size as well as dynamic nature of software increases, it is difficult to achieve and produce reliable and consistence software for end user. In addition to this, the researcher suggests conducting further research on a software which has a large dataset by applying various other computational intelligent techniques for software fault predictions. Again, the researcher suggests to design and develop a model which integrates with other computational intelligent techniques in order to establish estimation models that cable to estimate the reliability of software more consistency with least precision errors and cost.

Kumar et al. [3] used various machine learning approaches such as artificial neural network (ANN) which include backpropagation neural network (BPNN), radial Basis function network (RBFN), Elman network, support vector machines (SVM), cascade correlation neural networks (CCNN), decision tree (DTs), and fuzzy inference system (FIS) for the prediction of quality and error-free software developed on past history failures of software outputs, and SVM model depicted a better result while comparing the model applying ANNs, CCNN, DTs, and FIS in total datasets [10–12].

The researcher concluded that in order to make reliable software, the researcher should have large dataset and experimental investigation which are capable of validating the system by survey and experiment which are required in future research. Therefore, current research of software reliability forecasting applying computational intelligent learning approaches offers the direction for upcoming research in the area of software engineering in order to evaluate the influence of past and present failure of data for genuine software reliability prediction. Further they suggest for the new researcher, identical types of finding need to be established with large dataset and real-life scenario in various data sets to provide a comprehensive result across various institutions and make cost benefit analysis in order to determine software reliability prediction technique would be economically feasible in a real environment.

Pai et al. [8] propose support vector machine technique combined with genetic algorithms for predicting software reliability. Support vector machine is used for investigating a problem of nonlinear regression and time series which indicated that the behavior of software reliability prediction depends on time. Software reliability prediction commonly changes with time. These changes can be handled as a time series process. Predicting the variability of software reliability with time is challenging one. Finally, they suggest for the researcher that additional research should apply more efficient optimization algorithms to select the parameters of the support vector machine technique to forecast software reliability.

Karunanithi N. et al. [13] made an investigation of detailed study to exhibit the use of connectionist models or neural network model, which is applying in the software reliability prediction model. The approach of the researcher tries to identify several network models, training rules, and data representation methods. The researchers suggest that the connectionist models may adapt well across different data sets and show a better predictive accuracy, but various models have different analytical efficiency at various stages of software testing, and there is no common parametric model which can show accurate predictions in all software developments. Therefore, additional research should be conducted which considers all software development characteristic.

Cai et al. [14] proposed a fuzzy-based software reliability prediction models rather than probabilistic model. The authors claim that software reliability models are fuzzy in nature. Ho et al. mainly study intensively on connectionist models and their applicability to software reliability prediction. The findings of the researcher are that connectionist models are better as compared to traditional software reliability models.

Lou et al. [15] propose and discuss about the relevance of vector machine used for software reliability prediction. Yang et al. develop a hybrid model. The model uses data mining techniques and genetic algorithms for software reliability prediction. Authors have identified the utilization and applicability of auto-regressive integrated moving average technique and SVM for reliability prediction. Kumar et al. [3] propose the use of correlation neural networks, decision trees, and fuzzy inference system in their machine learning model to predict the reliability of software products.

Torrado et al. [16] focused on software failures over time. For this, the researcher used ML techniques like the Bayesian model alongside Gaussian processes. The proposed model uses a suitable multistep prediction approach. The model is used to predict long-term software failures.

Kulamala et al. [17] use numerical intelligence for software reliability prediction. They mentioned that for software reliability prediction, model development is very challenging. It is because of the dynamic and uncertain nature of the software. They conclude that software reliability prediction is an open problem and needs more research on it.

Ma et al. [18] use support vector regression for software reliability. The researcher gives a mathematical method to select the kernel function. Researcher uses root mean square error, mean absolute error, relative square root error, and relative absolute error results with reliability index parameters of the software for a correlation analysis.

Jabeen et al. [19] use a precision error iterative analysis method for predicting software reliability prediction. They compared their model with genetic algorithm. The output is measured using goodness-of-fit and predictive performance. They conclude that their model is versatile and universal in improving the performance of various software reliability growth models. Limitation of their research is not conducting large-scale comparison.

Tejaswini et al. [20] used machine learning approaches for software fault prediction. The paper lacks clarity about the research and how they carried out it. The researcher emphasizes on selecting a suitable software reliability growth, and prediction model in the software development is challenging for the developers. The idea is clear and accurate with regard to reliability measurement in software development life cycle.

Behera et al. [21] use chemical reaction optimization-based hybrid model for software reliability prediction. They further proposed that for improving accuracy of software reliability prediction, evolutionary techniques and higher-order neural network should be used.

Banga et al. [22] proposed a framework for software reliability prediction [36]. Software Failure and Reliability Assessment Tool (SFRAT) is proposed by Nagaraju et al. This tool having various software reliability growth models is made public as open-source tool.

Saraf et al. [23] proposed a model that is derived from a nonhomogeneous Poisson process (NHPP) based on a unified scheme for multi-release two stage fault detection or observation and correction or removal software reliability models.

4 Results

This portion states the outputs acquired from selected main investigation and incorporates specifics about the search outputs, a visualization of journal years and sources, and subsequent from this an outline of the quality evaluation output.

4.1 Explanation of Primary Studies

In this portion, we deliver a brief explanation of the selected main studies. We have selected 60 key studies, out of 250 studies which are developed on the quality queries, on prediction of software reliability using machine learning techniques. Most of the studies were extracted from open-source data sets.

4.1.1 Year of Publications and Source

When the researcher performs a systematic literature review analysis, the type of publication source and its publication year are the most important. In this systematic literature review, the publication years of selected primary studies lie between the year 2010 and 2020, and the following figure shows the numbers of studies

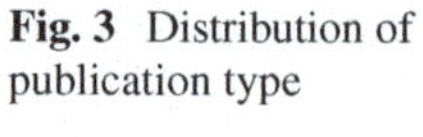

Fig. 3 Distribution of publication type

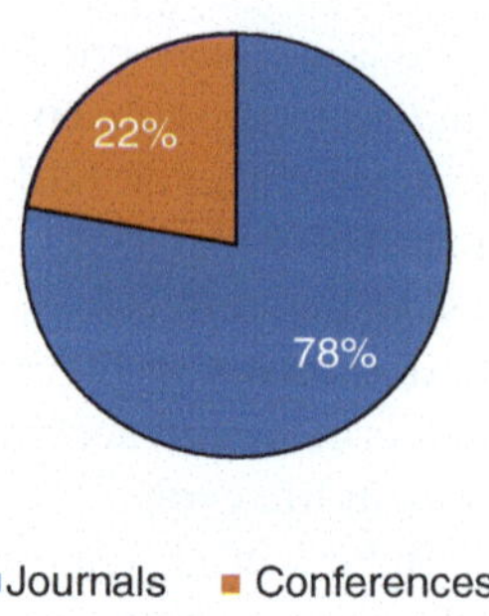

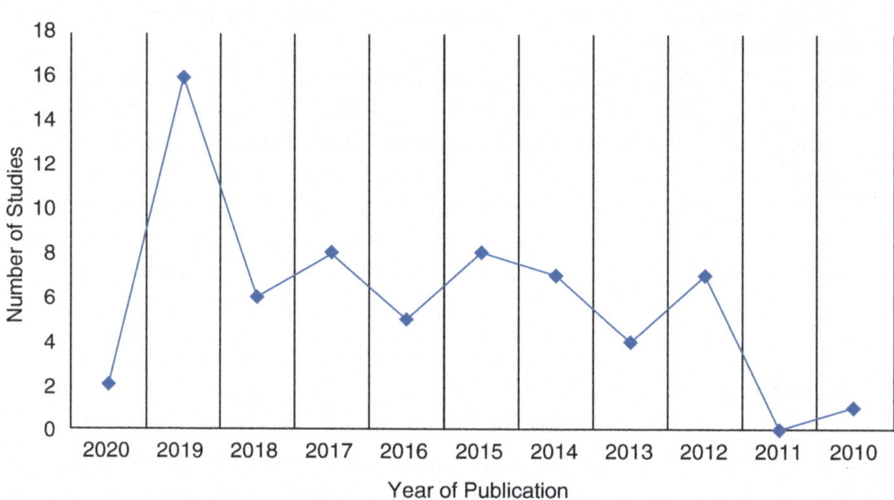

Fig. 4 Number of selected studies over the years

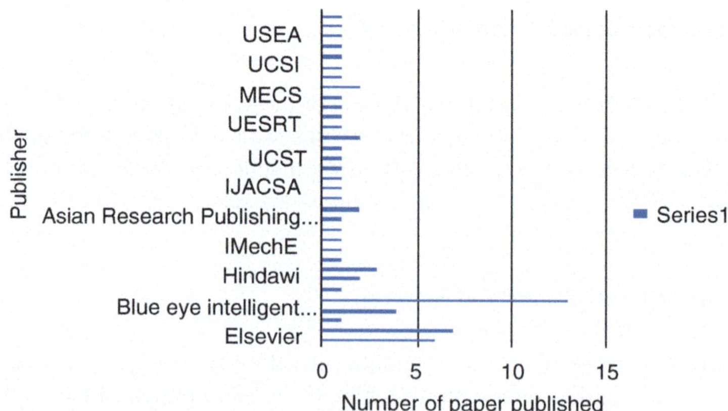

Fig. 5 The number of studies in each journal

published during these years. The major source of publications which is extracted from IEEE is as follows:

1. IEEE Xplore, ACM, Science Direct, Wiley Online Library, Springer, and so on. The sources of the publication come from two sources which is 78% of the studies come from journals and 22% of the studies come from conferences (Figs. 3, 4 and 5).

5 Conclusion

The review of this chapter will help the researchers to assess and evaluate available study done on software reliability prediction and particularly using machine learning. This review will give a clear picture of what the researcher has done on parameters, datasets, objective, methods, performance evaluation metrics, and experimental result perspectives. For conducting this review, we identified 60 primary studies which are relevant to the objective of our review literature. This study result shows that this topic is an open topic with many works needed to be done in terms of size of the software, language use in the software, and reliability models used for prediction. Soft computing and machine learning techniques are used, but the possibility of using more ML models needs to be explored.

References

1. Kitchenham, B., et al. (2009). Systematic literature reviews in software engineering – A systematic literature review. *Information and Software Technology, 51*(1), 7–15.
2. Jaiswal, A., & Malhotra, R. (2018). Software reliability prediction using machine learning techniques. *International Journal of System Assurance Engineering and Management, 9*(1), 230–244.
3. Kumar, P., & Singh, Y. (2012). An empirical study of software reliability prediction using machine learning techniques. *International Journal of System Assurance Engineering and Management, 3*(3), 194–208.
4. Srinivasan, K., & Fisher, D. (1995). Machine learning approaches to estimating software development effort. *IEEE Transactions on Software Engineering, 21*(2), 126–137.
5. Ali, A., Jawawi, D. N., Isa, M. A., & Babar, M. I. (2016). Technique for early reliability prediction of software components using behaviour models. *PLoS One, 11*(9).
6. Latha, D. H., & Premchand, P. (2018). Predicting software reliability using Ant Colony optimization technique with travelling salesman problem for software process-a literature survey. *International Journal on Recent and Innovation Trends in Computing and Communication, 6*(4), 106–112.
7. Aleem, S., Capretz, L. F., & Ahmed, F. (2015). Benchmarking machine learning technologies for software defect detection. *arXiv* preprint arXiv:1506.07563.
8. Pai, P. F. (2006). System reliability forecasting by support vector machines with genetic algorithms. *Mathematical and Computer Modelling, 43*(3–4), 262–274.

9. Ramasamy, S., & Lakshmanan, I. (2017). Machine learning approach for software reliability growth modeling with infinite testing effort function. *Mathematical Problems in Engineering, 2017.*
10. Martens, A., Koziolek, H., Becker, S., & Reussner, R. (2010, January). Automatically improve software architecture models for performance, reliability, and cost using evolutionary algorithms. In *Proceedings of the first joint WOSP/SIPEW international conference on performance engineering* (pp. 105–116). ACM.
11. Aljahdali, S. H., & El-Telbany, M. E. (2009, May). Software reliability prediction using multi-objective genetic algorithm. In *IEEE/ACS International Conference on Computer Systems and Applications, 2009 (AICCSA 2009).* (pp. 293–300). IEEE.
12. Pati, J., & Shukla, K. K. (2015, February). A hybrid technique for software reliability prediction. In *Proceedings of the 8th India software engineering conference* (pp. 139–146). ACM.
13. Karunanithi, N., Whitley, D., & Malaiya, Y. (1992). Prediction of software reliability using connectionist models. *IEEE Transactions on Software Engineering, 18*(7), 563–574.
14. Cai, Y. K., Wen, Y. C., & Zhang, L. M. (1991). A critical review on software reliability modeling. *Reliability Engineering and System Safety, 32*(3), 357–371.
15. Lou, J., Jiang, Y., Shen, Q., Shen, Z., Wang, Z., & Wang, R. (2016). Software reliability prediction via relevance vector regression. *Neurocomputing, 186*, 66–73.
16. Torrado, N., Wiper, M. P., & Lillo, R. E. (2013). Software reliability modeling with software metrics data via gaussian processes. *IEEE Transactions on Software Engineering, 39*(8), 1179–1186.
17. Kulamala, V. K., Maru, A., Singla, Y., & Mohapatra, D. P. (2018, December). Predicting software reliability using computational intelligence techniques: A review. In *2018 international conference on information technology (ICIT)* (pp. 114–119). IEEE.
18. Ma, Z. Y., Zhang, W., Wang, J. P., Liu, F. S., Han, K., & Gao, F. (2018, December). Research on a method of software reliability prediction model. In *Proceedings of the 2018 2nd international conference on computer science and artificial intelligence* (pp. 163–167). ACM.
19. Jabeen, G., Luo, P., & Afzal, W. (2019). An improved software reliability prediction model by using high precision error iterative analysis method. *Software Testing, Verification and Reliability*, e1710.
20. Tejaswini, P., Varsha, K., Yasaswini, P., & Yalamanchili, S. (2019, September). Software defect prediction using machine learning. *International Journal of Recent Technology and Engineering (IJRTE), 8*(2), S11. ISSN: 2277-3878.
21. Behera, A. K., Nayak, S. C., Dash, C. S. K., Dehuri, S., & Panda, M. (2019). Improving software reliability prediction accuracy using CRO-based FLANN. In *Innovations in computer science and engineering* (pp. 213–220). Springer.
22. Banga, M., Bansal, A., & Singh, A. (2019, April). Implementation of machine learning techniques in software reliability: A framework. In *2019 International Conference on Automation, Computational and Technology Management* (ICACTM) (pp. 241–245). IEEE.
23. Saraf, I., & Iqbal, J. (2019). Generalized multi release modelling of software reliability growth models from the perspective of two types of imperfect debugging and change point. *Quality and Reliability Engineering International.*

Evolutionary Algorithms for Path Coverage Test Data Generation and Optimization: A Review

Dharashree Rath, Swarnalipsa Parida, Deepti Bala Mishra, and Sonali Pradhan

1 Introduction

Software testing is a process to analyze whether the performed results match with the expected results or not and to produce catastrophic bugs-free software module [1]. Nowadays, almost in every sector, the role of software is very crucial, and the requirement of software technology plays a vital role in our society. Though software becomes intricate, the demand of good software is rising in the size, and emphasis is given on the quality of software product [2]. So, software testing is mandatory to find a quality software, and it is an important phase in every Software Development Life Cycle (SDLC) shown in Fig. 1. To increase the quality of a software product, software testing is used, and this is considered as a vital role in software development phase [3]. In software development, software testing is considered as a costly and time-consuming process. Also, it helps to improve the confidence of the consumer. The reason of software testing is to develop a reliable, trustworthy, and robust software product [4]. To perform testing operation the software modules are divided into some small and isolated parts which are referred as test cases. Then the test cases provide some input data values to the test system to analyze the errors [5]. Every year, software company recompenses an enormous loss of about $500 billion per year due to reduction in software product quality, and the reason of software failure is the negligence of software company, and to eradicate the failure, software testing is taken, and they get some positive results [6, 7].

D. Rath · S. Parida · D. B. Mishra (✉)
GITA Autonomous College, Bhubaneswar, India

S. Pradhan
C. V. Raman Global University, Bhubaneswar, India

M. Khari et al. (eds.), *Optimization of Automated Software Testing Using Meta-Heuristic Techniques*, EAI/Springer Innovations in Communication and Computing, https://doi.org/10.1007/978-3-031-07297-0_7

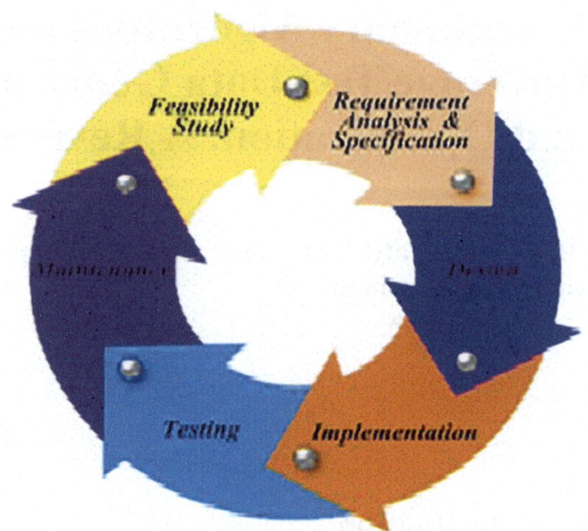

Fig. 1 Phases of SDLC

Software testing is classified as random-based software testing and search-based software testing [8]:

- RBST.
- SBST.

For test data generation, RBST is considered the simplest technique, that is, here the random inputs are given as input data for execution, and then it examines the results that whether these are satisfied or not, but the constraint satisfying probability is low for the programs which are already tested. But the main drawback of RBST is that in some cases the test data does not match with the target test data, and this situation is known as critical data [9].

The SBST technique is considered the most popular technique as many software industries are using it for solving the optimization problem and search-based testing problems by using some heuristic algorithm. There is some metaheuristic algorithm, namely, genetic algorithm (GA), particle swarm optimization (PSO), artificial bee colony optimization (ABCO), ant colony optimization (ACO), etc. [10].

The next part of this chapter is organized as follows: Section 2 describes the basic concept of testing and the working principles of some testing techniques. Section 3 represents the review parts as we took some papers and review their works and experiments on different evolutionary algorithms using in the testing techniques. Lastly, Section 4 describes the conclusion part of our review and some observation from our systematic literature studies.

2 Basic Concepts

Here a brief description about the basic concept of some testing approaches of those which are used in our work is described.

2.1 *Testing Levels*

Generally, a software testing goes under some sub-testing techniques [11–13], that is:

- Unit testing.
- Integration testing.
- System testing.
- Acceptance testing.
- Regression testing.

2.1.1 Unit Testing

Unit testing is considered as small testing of some modules or some parts of any given software project. After a software is designed and coded, unit testing takes place. Here the unit test cases are introduced and ready to design and then tested. The module is divided into two parts, that is, stub and driver. The stub is considered as a dummy and simplified procedure that is similar to the given I/O parameters.

In unit testing, stubs are referred to as "called program" function which is used in top-down approach, and after the completion of the top level, the lowest level is tested.

And drivers are referred to "called program" function which is used in bottom-up approach, and after the completion of bottom level, top level is tested.

2.1.2 Integration Testing

After completion of unit testing, it combined as a group and tested logically that means data flow between the dependent module is tested here. In integration testing, the integrated modules are combined and tested whether some faults detect after combination of modules.

Different types of integration testing are introduced, namely, Big Bang integration, top-down integration, bottom-up integration, mixed integration, etc.

2.1.3 System Testing

The complete system is evaluated and tested when all modules are combined.

The main aim of this testing is to satisfy the user requirement and specification of the system. System testing is used to check the errors from the integrated modules and whole system.

There are various testing techniques used like performance, load, stress, and scalability. Basically, system testing comes under some types that are listed as follows:

(a) Set the testing environment.
(b) Generate test cases.
(c) Test data generation.
(d) Execution of test cases.
(e) Reporting of defects.

2.1.4 Acceptance Testing

In acceptance testing acceptability of system is tested to check that it fulfils the business requirement of customer or not. The main goal of this testing is to check whether the software products are ready to deliver or not.

There are two types of testing, namely, alpha testing and beta testing:

- Alpha testing: Performed at developer side.
- Beta testing: Performed at customer side.

2.1.5 Regression Testing

Regression testing is determined to check the changes in new functionality, but it does not affect the old functionality and also confirms that the previous functionality works perfectly with the new functionality.

When it comes to the maintenance of software, bug fix may be needed, and the work of regression testing is to check that the newly added feature has been no inauspicious effect on the existing working software.

If a modified software module fails or when a new module is used with previous unchanged module and causes error in unchanged module by creating complication, then the system under test (SUT) is said to regress.

2.2 Black Box Testing

In black box testing, the functionality of software specification is examined, and the tester is not aware of the internal and structural design of the software product [14].

This testing mainly emphasizes on both the function and behavior of software. The tester can perform specification using the following techniques:

- Boundary value analysis.
- Equivalence partitioning.
- Graph based.

- Transition state.
- Decision table.
- Similarity testing.

2.3 *White Box Testing*

In white box testing, the main focus is on structural and internal design. As tester is known about the code, so it is known as clear boxing [15]. When testers are testing a program using white box testing, it should be followed by two basic steps, namely:

(a) **Understand the Source Code:** At the beginning of testing, tester should learn and understand the program code and then use white box testing for internal and structural design. The tester is highly responsible for issue and also checks the prevention from the hackers and attackers.
(b) **Create Test Cases and Execute:** In this case, the tester should create test cases for each program and then execute the program step by step.

2.3.1 Fault-Based Testing

To eradicate the specific category of failure, fault-based testing is used such as mutation testing.

2.3.2 Coverage-Based Testing

Here, testing is done on the basis of certain criteria, and requirement of this testing is to choose the test cases which are equivalent to the criteria and different coverage factor, namely, path coverage, condition coverage, branch coverage, statement coverage, etc. that are described as follows [15]:

(a) **Statement Coverage**: In this testing procedure, the application which comes under test is tested at least once, and when the procedure will start, we should check the probable error in the application.

    The main drawback of this technique is that the statements are always checked once and decided that it functions properly.
(b) **Branch Coverage**: In this testing, test cases are designed to create branch condition in the program and all and also presume that it is true or false. It is stronger than statement-based testing.
(c) **Condition Coverage**: Here, test cases are designed on the basis to make a composite conditional statement and assume that it is true or not and then execute all the expression and conclude the outcomes.
(d) **Path Coverage**: In this testing, test cases are designed in the manner that all linearly independent paths are executed at least once.

2.4 Path Testing

Path testing is a structural testing technique that was introduced by How den in 1976, designed in such a way that all selected paths are executed through a computer program. Here a set of test data traversed all the logical paths and executed for accurate results and branch coverage. So, it is considered as a secure coverage technique than other testing process [16].

It uses two sub-steps, that is, test data generation and target path generation for covering the target path. In this technique, all linearly independent executable paths are covered to find all the underlying faults inside every piece of code, and some cyclomatic complexity methods are used to determine that linearly independent executable paths. The cyclomatic complexity (CC) was designed by McCabe, and it provides linear independent path and the upper bound value [12, 16]. The cyclomatic complexity of a program can be represented by Eq. (1).

$$V(G) = E - N + 2 \tag{1}$$

where E is the total number of Edges, N is the number of nodes present in the graph G.

2.4.1 Critical Path

In critical path testing, no test data is generated, and it takes time to search the test data. For generating a test suite which is optimized and creates challenges for others, that means for future, it is considered as a meaningless searching process [15–17]. There are different heuristic methods like GA, ACO, PSO, and ABC that are used to generate test data accompanying with critical path.

2.4.2 Control Flow Graph

Control flow graph is used to determine the logical path. Here the flow sequence of programs is represented by nodes and edges [15, 18]. Nodes are used to generate decision, and edges are used to control the statement, and edges are used to control the statement. Different approaches are used to determine linearly independent paths with the help of path testing. For all possible paths, test cases are executed and determined with 100% results of statement and branch coverage [18].

Hence, from Fig. 2, the linearly independent paths are as follows:

1→6
1→2→3→5→6
1→2→4→5→6

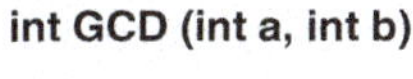


```
int GCD (int a, int b)
  1. while (a! =b)
  2. { if(a>b) then
  3. a=a-b;
  4. else b=b-a;
  5. }
  6. return a;
```

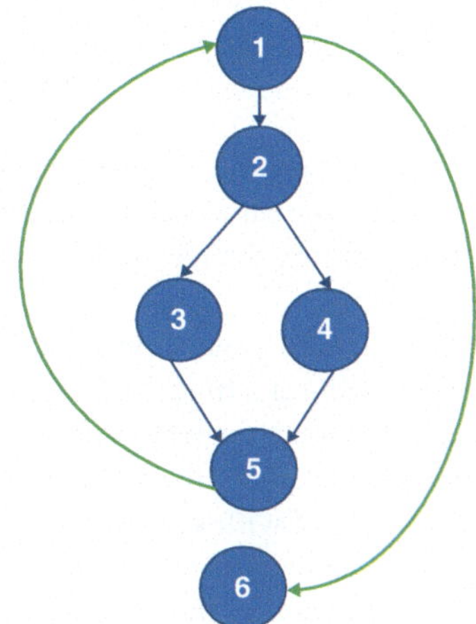

Fig. 2 CFG for GCD program

For generation test data from multiple paths, path testing can be formulated as an optimization problem which is explained in equation [19].

Problem Let S is the SUT.

Here "I" is the input domain.

$p = \{p_1, p_2, p_3, p_4, \ldots\}$ is its target path set for 100% path coverage.

We need to find a set of test data $X = \{x_1, x_2, x_3, x_4, \ldots\}$ where $x_i \in I$ and $f(x)$ is a function which can be mapped tox_i to cover a path p_i in S, defined in Eq. (2):

$$f(x) = \frac{p(x_i)}{p_i} \tag{2}$$

So, x is desired test data for traversing path p_i if $f(x) = 1$.

3 Related Work

In this section, we discuss some related state-of-the-art works on the frequent use of EAs for optimizing the test data for path coverage-based testing to find the efficient path. The works are briefly described as follows.

3.1 Test Case Generation and Optimization Using GA

***Mansour* et al.** [12] proposed two algorithms, namely, simulated algorithm (SA) and genetic algorithm (GA) to generate the test cases which execute the selected paths of a program. These two algorithms include integer value and real value test cases. And the outcome shows that these two algorithms give faster results than Kore's algorithm. Joining SA and KA, execution time is decreasing and higher success rate is achieved.

***Zhang* et al.** [20] (2014) have presented a novel method to generate test data for covering multiple paths at a time. The proposed method not only can cover multiple paths at a time but also can detect faults present in the SUT. A weighted GA-based model is constructed to solve their multi-objective optimization problem. The proposed method is applied with several real-world programs like bubble sort and programs of Siemen's suite as print_tokens.c, replace.c, schedule.c, tcas.c, etc. Their reported results confirm that the proposed technique can generate test data for traversing target paths as well as detecting maximum faults lying in the SUT.

***Chen* et al.** [21] proposed a technique to optimize the backpropagation (BP) neural network using improved adaptive genetic algorithm, that is, named as IAGA-BP, which facilitates the use of Wi-Fi fingerprint data for positioning an object. To enhance the biases and weights of the BP neural network, the selection, crossover, and mutation operation of genetic-based algorithm are used. Each generation population will be arranged on the basis of capacity from high to low, and 20% population will come into the next level directly, where 20% population will be cancelled and then the rest 80% is preferred by roulette algorithm. Lastly to improve the accuracy of Wi-Fi positioning and to generate the convergence speed of neural network, the IAGA-BP model can be used.

***Deepti* et al.** [22] have given an extensive study on different white box testing technique using GA. The authors have discussed about the path coverage-based testing as well as the mutation testing. In their work every unit of software under test (SUT) is tested by the developers, and GA has been used to generate an optimized test data to cover the critical path as well as kill the mutants present in the SUT. A complete scenario of the previous work can be drawn by their research work.

***Robert* et al.** [23] presented a color image segmentation which used the graph cut method, and this also used two constraints like color constraint and gradient constraint by minimizing the weighted energy function. Here, the researchers found the results by taking input images where a number of segments are presented in a single image and the algorithm and by applying genetic algorithm, that is, crossover and mutation operators applied to various segments of the subgraphs which are obtained by the graph of graph cut method. At last, the optimized method OGcut has been applied which results in some non-overlapping segments. Some constraints are also added in the proposed method to enhance the score of overlapping.

***Latiu* et al.** [24] took three different evolutionary algorithms like GA, PSO, and SA which help to generate test data automatically and take some comparison among them. They compared the PSO and SA (simulated annealing) with the GA by correlating the branch distance metrics and approximation level.

Huang **et al.** [25] presented a noble technique using the operator self-activity feedback (SAF) and Gaussian mutation (G) to enhance the performance of PSO by substituting the weight inertia. The improved PSO algorithm provides a better performance in multi-path test case generation.

Han **et al.** [15] designed a modified multi-path particle swarm optimization (MMPPSO) to generate test data for multi-path. Authors involved various fitness functions that are used to find the best position (pbest) and global position (gbest). A branch distance function was designed to calculate the weight summation of single path fitness function.

Sun **et al.** [26] designed an algorithm to improve both position and particle velocity by modifying randomness. To enhance the algorithm evolution, a converge analysis is used to find the inertia weight that provides the factor that has been accelerated. Also, a random sampling strategy is used to determine the sample of parameter setting and also modify randomness. To retain the convergence accuracy and also enhance the rate of convergence, PSO algorithm is used.

Digehsara **et al.** [27] presented an algorithm to overcome the problems in PSO, that is, selection of initial particle in PSO randomly, unable to balance search space. To handle this problem, Halton-PSO was designed. Halton-PSO is union of initial PSO and Halton sequence. Eleven benchmark functions and seven nonlinear engineering problem are used for enhancement of PSO.

Biswas **et al.** [28] presented ACO-based algorithm to generate the optimal path for a complete path coverage. This method provides a test data sequence that are used as inputs to the generated path in the domain. This method assured that the software path coverage has minimum redundancy.

Mann **et al.** [29] developed a path-prioritization-ant-colony optimization (PP-ACO), that is, used to find all decision to decision path using CFG. Also, it focuses to prioritize the optimal path and extract the path sequence using DD graph. According to the path strength, path sequence was determined and also prioritizes the best optimal path.

Wang **et al.** [30] designed an algorithm for analysis of multiple travelling salesman problem (MTSP), which relied on improved ant colony optimization. This paper is based on logistic factor distribution and vehicle capacity of every salesman. Union of MTSP and 1-MST is used in improved ACO. To find optimal path and efficient ACO algorithm, the combination of minimum spanning I-tree with ACO is used. It states that though the application can't assist the logistic distribution, yet it enhances the relief work of earthquake and also makes the military operation flexible.

Xiao **et al.** [31] presented an algorithm to enhance the consumption of energy-efficient clustering in UWSNs which relied on improved ACO algorithm. Two nodes, namely, cluster head node (CHN) and cluster member node (CMN), are used to optimize the residual energy of nodes. CHN collects data, that is, it is transmitted by CMN and sent them to the sink node using multiple hops. In this paper, the development of heuristic information, pheromone update mechanism, and ant searching scope are used to improve ACO algorithm.

***Lam* et al.** [32] proposed an automatic generation of feasible independent paths and test suite optimization using artificial bee colony (ABC). In this method, ABC merges both gbest (global best) which are created by scout bees and pbest (local best) method done by employed bees and onlooker bees. These three bees generate feasible independent paths and also make the optimization process faster.

***Khari* et al.** [33] designed two methods, that is, test suite generation and test suite optimization. Different suite generation methods were found such as robustness testing, worst-case testing, boundary value testing, random testing, and robust-worst-case testing. Using artificial bee colony algorithm, the generation of test suite was optimized to find the appropriate fitness level. This technique provides maximum path coverage and less redundancy than other algorithm.

***Sun* et al.** [34] designed a technique to enhance the improvement of artificial bee colony (ABC) algorithm along with random artificial bee colony (RABC) which are used to explore the solution, and also the search scope method called tournament selection strategy that is used to enhance the evolutionary process by improving diversity population outcome of the analysis determines that RABC is accurate and convergence speed is greater than other algorithm.

***Malhotra* et al.** [35] took some utility-based algorithms like GA, ACO, and ABCO and evaluated their various factors by taking test suites. The different factors like number of covered paths and iteration number are taken, and test case number is generated and generates the test suite time. At last, they observed that the ABCO algorithm is efficient to generate test suites than the other two algorithms.

***Huang* et al.** [36] developed an algorithm in 2018 for enhancing (ATCG-PC) automated test case generation based on path coverage using fog computing. To decrease the problem of classical differential evolution (DE), a test case path relationship matrix (PRM) is used, that is, RP-DE. Also, the fog computing uses programs like, iFogSim toolkit, which provides infeasible paths. The RP-DE is developed by involving programs from iFogSim toolkit, and also it provides higher path coverage and lower test cases.

***Guo* et al.** [37] constructed an algorithm using preventive maintenance period optimization mode (PMPOM) to receive an optimal maintenance period. A hybrid algorithm, namely, PSOCS, is generated using the merits of Cuckoo search (CS) and particle swarm optimization (PSO) algorithm. The reported result shows that the proposed PSOCS is more efficient in convergence speed than the other previous methods. The convergence speed is very high during solving the PMPOM problem.

***Utama* et al.** [38] proposed an algorithm using hybrid butterfly optimization algorithm (HBOA) to reduce the costs on green vehicle routing problem (GVRB). Both butterfly optimization algorithm (BOA) and tabu search (TS) algorithm are flip and swap method used to inspire HBOA algorithm. And it is concluded that HBOA is used to minimize the cost distribution than other algorithms.

4 Conclusion

In this chapter, some related works on path coverage-based software testing have been reviewed. Different types of static methods as well as dynamic methods have already been used to generate and optimize the test data for path testing. For the optimization purpose so many optimization techniques are also used, to cover the critical path present in the software. It is also observed that to achieve highest path coverage for more complex software, more efficient method is needed since the main issue is the presence of critical paths in the software, for which it is very difficult to find an optimal test suite with maximum coverage. It is also observed that the process of detecting and generating the test data to cover a critical path is the most challenging issue during software path testing.

References

1. Chauhan, N. (2010). *Software testing: Principles and practices*. Oxford University Press.
2. Mall, R. (2018). *Fundamentals of software engineering*. PHI Learning Pvt. Ltd.
3. Zhu, Z., Xu, X., & Jiao, L. (2017, June). Improved evolutionary generation of test data for multiple paths in search-based software testing. In *2017 IEEE Congress on Evolutionary Computation (CEC)* (pp. 612–620). IEEE.
4. Srivastava, P. R., & Kim, T. H. (2009). Application of genetic algorithm in software testing. *International Journal of Software Engineering and its Applications, 3*(4), 87–96.
5. Alshraideh, M., Mahafzah, B. A., & Al-Sharaeh, S. (2011). A multiple-population genetic algorithm for branch coverage test data generation. *Software Quality Journal, 19*(3), 489–513.
6. Mishra, D. B., Bilgaiyan, S., Mishra, R., Acharya, A. A., & Mishra, S. (2017). A review of random test case generation using genetic algorithm. *Indian Journal of Science and Technology, 10*(30).
7. Bhuyan, M. K., Mohapatra, D. P., & Sethi, S. (2016). Software reliability prediction using fuzzy min-max algorithm and recurrent neural network approach. *International Journal of Electrical and Computer Engineering (IJECE), 6*(4), 1929–1938.
8. Manikumar, T., Kumar, A. J. S., & Maruthamuthu, R. (2016). Automated test data generation for branch testing using incremental genetic algorithm. *Sādhanā, 41*(9), 959–976.
9. Sharma, A., Rishon, P., & Aggarwal, A. (2016). Software testing using genetic algorithms. *International Journal of Computer Science and Engineering Survey (IJCSES), 7*(2), 21–33.
10. Torkamani, M. A. (2014). Metric suite to evaluate reusability of software product line. *International Journal of Electrical and Computer Engineering (IJECE), 4*(2), 285–294.
11. Khari, M., & Kumar, P. (2017). An extensive evaluation of search-based software testing: A review. *Soft Computing*, 1–14.
12. Mansour, N., & Salame, M. (2004). Data generation for path testing. *Software Quality Journal, 12*(2), 121–136.
13. Mishra, D.B., Mishra, R., Das, K.N., & Acharya, A.A. (2017). A systematic review of software testing using evolutionary techniques. In *Proceedings of sixth international conference on soft computing for problem solving* (pp. 174–184). Springer.
14. Hermadi, I., Lokan, C., & Sarker, R. (2010, December). Genetic algorithm based path testing: Challenges and key parameters. In *2010 second World Congress on Software Engineering (WCSE)* (Vol. 2, pp. 241–244). IEEE.

15. Han, X., Lei, H., & Wang, Y. S. (2016). Multiple paths test data generation based on particle swarm optimization. *IET Software, 11*(2), 41–47.
16. Garg, D., & Garg, P. (2015). Basis path testing using SGA & HGA with ExLB fitness function. *Procedia Computer Science, 70*, 593–602.
17. Shimin, L., & Zhangang, W. (2011). Genetic algorithm and its application in the path-oriented test data automatic generation. *Procedia Engineering, 15*, 1186–1190.
18. Boopathi, M., Sujatha, R., Kumar, C.S., & Narasimman, S. (2014, October). The mathematics of software testing using genetic algorithm. In *2014 3rd International Conference on Reliability, Infocom Technologies and Optimization (ICRITO) (Trends and Future Directions)* (pp. 1–6). IEEE.
19. Shahbazi, A., & Miller, J. (2016). Black-box string test case generation through a multi-objective optimization. *IEEE Transactions on Software Engineering, 42*(4), 361–378.
20. Zhang, Y., & Gong, D. (2014). Generating test data for both paths coverage and faults detection using genetic algorithms: Multi-path case. *Frontiers of Computer Science, 8*(5), 726–740.
21. Chen, Y., Zhang, J., Liu, Y., Zhao, S., Zhou, S., & Chen, J. (2020). Research on the prediction method of ultimate bearing capacity of PBL based on IAGA-BPNN algorithm. *IEEE Access, 8*, 179141–179155.
22. Mishra, D.B., Mishra, R., Acharya, A.A., & Das, K.N. (2019). Test data generation for mutation testing using genetic algorithm. In *Soft computing for problem solving* (pp. 857–867). Springer.
23. Robert Singh, A., & Suganya, A. (2020, January). *Optimized Graph cut color image segmentation using genetic algorithm with weighted constraints (OGcut).*
24. Latiu, G.I., Cret, O.A., & Vacariu, L. (2012, September). Automatic test data generation for software path testing using evolutionary algorithms. In *2012 third international conference on Emerging Intelligent Data and Web Technologies (EIDWT)* (pp. 1–8). IEEE.
25. Huang, M., Zhang, C., & Liang, X. (2014, December). Software test cases generation based on improved particle swarm optimization. In *2014 2nd International Conference on Information Technology and Electronic Commerce (ICITEC)* (pp. 52–55). IEEE.
26. Sun, L., Song, X., & Chen, T. (2019). An improved convergence particle swarm optimization algorithm with random sampling of control parameters. *Journal of Control Science and Engineering, 2019*.
27. Digehsara, P. A., Chegini, S. N., Bagheri, A., & Roknsaraei, M. P. (2020). An improved particle swarm optimization based on the reinforcement of the population initialization phase by scrambled Halton sequence. *Cogent Engineering, 7*(1), 1737383.
28. Biswas, S., Kaiser, M.S., & Mamun, S.A. (2015, May). Applying Ant Colony optimization in software testing to generate prioritized optimal path and test data. In *2015 International Conference on Electrical Engineering and Information Communication Technology (ICEEICT)* (pp. 1–6). IEEE.
29. Mann, M. (2015). Generating and prioritizing optimal paths using ant colony optimization. *Computational Ecology and Software, 5*(1), 1.
30. Wang, M., Ma, T., Li, G., Zhai, X., & Qiao, S. (2020). Ant Colony optimization with an improved pheromone model for solving MTSP with capacity and time window constraint. *IEEE Access, 8*, 106872–106879.
31. Xiao, X., & Huang, H. (2020). A clustering routing algorithm based on improved Ant Colony optimization algorithms for underwater wireless sensor networks. *Algorithms, 13*(10), 250.
32. Lam, S. S. B., Raju, M. H. P., Ch, S., & Srivastav, P. R. (2012). Automated generation of independent paths and test suite optimization using artificial bee colony. *Procedia Engineering, 30*, 191–200.
33. Khari, M., Kumar, P., Burgos, D., & Crespo, R. G. (2017). Optimized test suites for automated testing using different optimization techniques. *Soft Computing*, 1–12.
34. Sun, L., Chen, T., & Zhang, Q. (2018). An artificial bee colony algorithm with random location updating. *Scientific Programming, 2018*.

35. Malhotra, R., & Kumar, N. (2016, September). Automatic test data generator: A tool based on search-based techniques. In *2016 5th International Conference on Reliability, Infocom Technologies and Optimization (Trends and Future Directions) (ICRITO)* (pp. 570–576). IEEE.
36. Huang, H., Liu, F., Yang, Z., & Hao, Z. (2018). Automated test case generation based on differential evolution with relationship matrix for iFogSim toolkit. *IEEE Transactions on Industrial Informatics, 14*(11), 5005–5016.
37. Guo, J., Sun, Z., Tang, H., Jia, X., Wang, S., Yan, X., Ye, G., & Wu, G. (2016). Hybrid optimization algorithm of particle swarm optimization and cuckoo search for preventive maintenance period optimization. *Discrete Dynamics in Nature and Society, 2016.*
38. Utama, D. M., Widodo, D. S., Ibrahim, M. F., & Dewi, S. K. (2020). A new hybrid butterfly optimization algorithm for green vehicle routing problem. *Journal of Advanced Transportation, 2020.*

A Survey on Applications, Challenges, and Meta-Heuristic-Based Solutions in Wireless Sensor Network

Neha Sharma and Vishal Gupta

1 Introduction

A wireless sensor network (WSN) helps industries to improve their productivity and reliability at low cost and hence used a lot in industries. A wireless sensor network is a wireless network consisting of spatially distributed autonomous devices that use sensors to monitor physical or environmental conditions [1, 2]. The WSN is a collection of a low-powered and small-sized sensor nodes [3], and these devices are autonomous in nature, and due to this feature, it can be installed in different environments. This feature results in numerous applications of WSN in real life. WSN is a field which is used for public as well as military purpose on large scale [4]. Applications like health care monitoring, water waste management, land irrigation, biodiversity mapping, agriculture, natural calamities, and normal household work are to be named of few, where sensor is used (IoT is a major application of WSN, used by general public day to day, to complete their basic works).

To accomplish the abovementioned tasks, one needs to handle the different issues of the WSN, namely, power management, node deployment, scalability, connectivity, coverage, security, routing efficiency, and localization problem. Meta-heuristic is a higher-level technique to generate or select an optimal solution of a problem or a better solution than others. Meta-heuristic is also used to optimize the solutions in WSN. For different issues of WSN, different meta-heuristic algorithms are applied. Numerous researchers have worked in the field of meta-heuristic, and they have

N. Sharma (✉)
AIACT&R, GGSIPU, Delhi, India

Department of IT ADGITM, GGSIPU, Delhi, India
e-mail: neha.sharma@adgitmdelhi.ac.in

V. Gupta
Department of CSE, NSUT East Campus, Formerly AIACT&R, Delhi, India
e-mail: vishal.gupta@aiactr.ac.in

M. Khari et al. (eds.), *Optimization of Automated Software Testing Using Meta-Heuristic Techniques*, EAI/Springer Innovations in Communication and Computing, https://doi.org/10.1007/978-3-031-07297-0_8

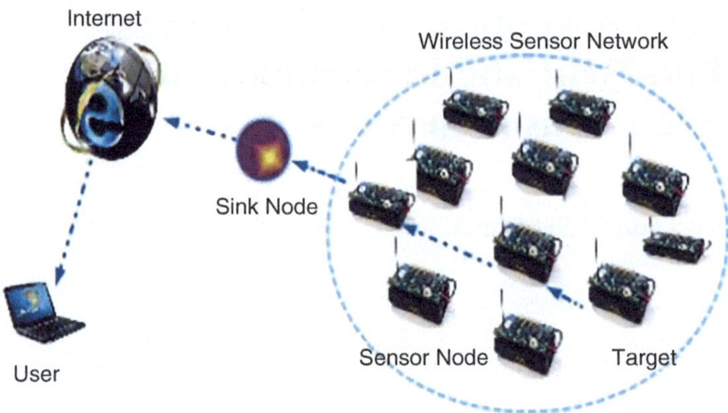

Fig. 1 Wireless sensor network [5]

suggested that the use of meta-heuristic in the field of WSN can result in more efficient and time-saving systems. Use of meta-heuristic helps in minimizing error in localization algorithms which are clustering based and helps in energy efficiency and routing efficiency. Fig. 1 shows the basic design of WSN.

Metaheuristic is amazing optimization technique, which can be used in number of real-world applications. Metaheuristic can be used in software testing automation. In the field of software, software coverage, test cases, generating test data, energy function improvement, elimination of explicit states, etc. can be optimized using meta-heuristic. This paper is divided into different sections: Section 2 consists of research methodology. Section 3 consists of different types of WSN. Section 4 analyzes the major application areas of WSN and its challenges. Section 5 focuses on meta-heuristic. Section 6 comprises of a literature review on meta-heuristic techniques used in WSN. And at last, Sect. 7 provides an informative conclusion of the paper.

2 Research Methodology

As the nature of research in wireless sensor network is very difficult to incarcerate to define disciplines, the related materials are scattered across various journals. The major challenges are localization problem, coverage and deployment, energy and power management, and routing efficiency in wireless sensor network.

Meta-heuristic can be used in solving challenges of WSN. Many researchers have collaborated WSN with meta-heuristic, and results suggest that solutions provided by these techniques are more efficient. These problems are the common for research in wireless sensor network. Accordingly, the following online journal databases were searched to provide a complete bibliography of the academic literature

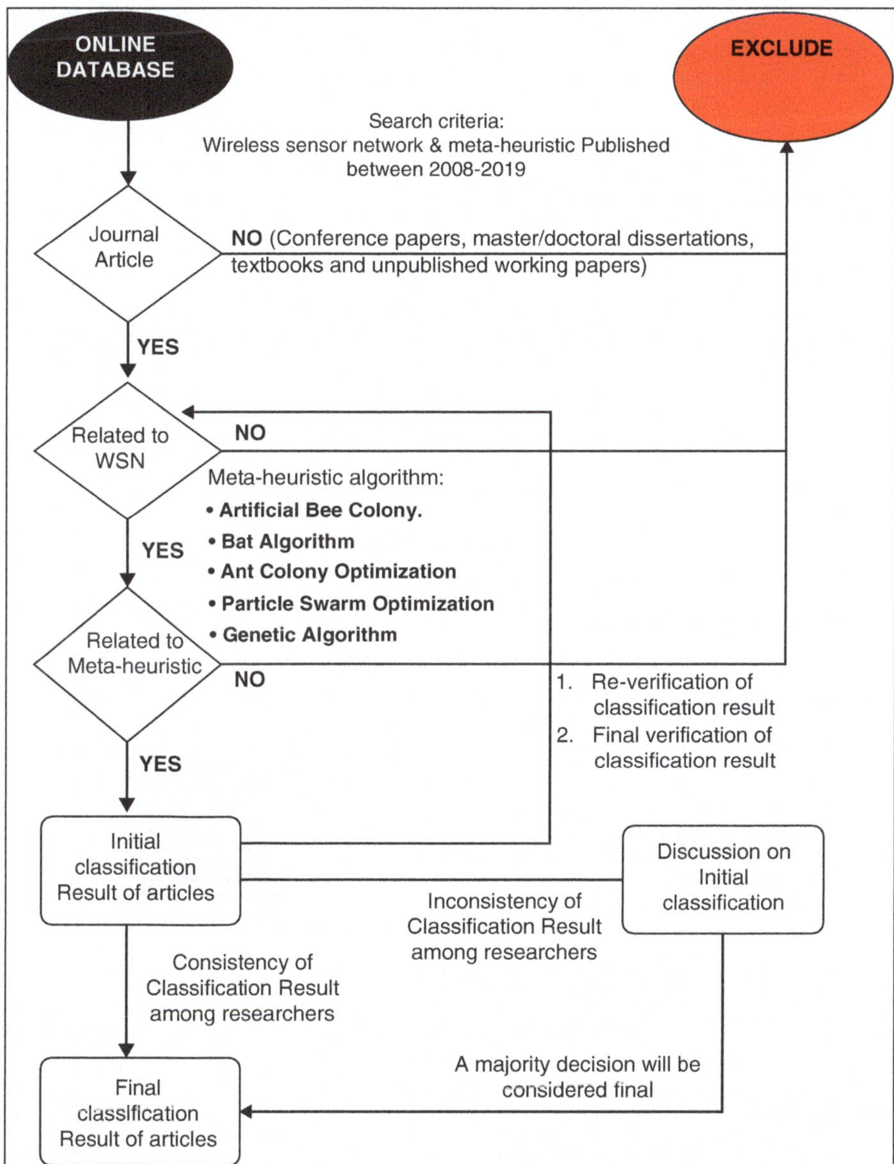

Fig. 2 Research methodology

on wireless sensor network and meta-heuristic. Figure 2 depicts the research methodology used for the survey.

- ACM digital library database
- Springer
- IGI Global

- Wiley
- Science Direct
- IEEE Transaction

3 Different Types of WSN

WSN are installed in different places, according to the requirement, like on land, underground, and underwater. Depending upon their deployment environment, they face different issues. In the current scenario, we are dealing with five categories of WSN, as shown in Fig. 3.

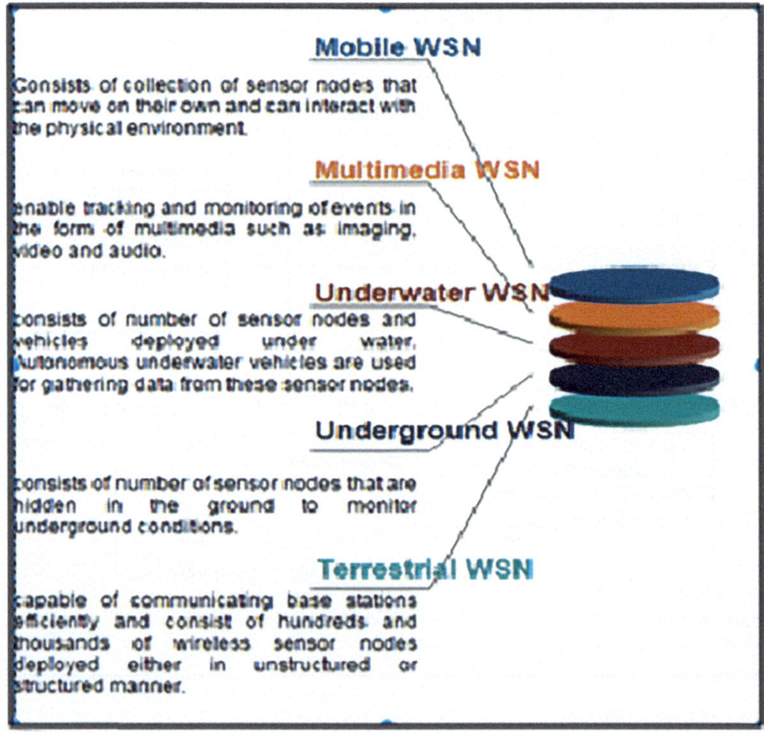

Fig. 3 Types of WSN

3.1 Mobile WSN

In mobile WSN, sensor nodes can interact with the physical environment and can move on their own [6]. It is a collection of nodes, which can sense the environment, communicate, compute information, and can also relocate. Since sensor nodes are mobile, only dynamic routing algorithms can be applied here. Examples of this category of WSN are tracking and searching and military surveillance. The coverage and connectivity provided by mobile WSN are better than static WSN [7].

3.2 Multimedia WSN

Multimedia WSN tracks and monitors events in the form of images, video, or audio form. Camera and microphone are equipped; low-cost sensor nodes are installed in a way to provide assured coverage [8]. But few problems are associated with this, and as multimedia form is used, high bandwidth is required, and quality of service and use of good compression techniques to smoothen the transfer and data processing are other issues.

3.3 Underwater WSN

In underwater WSN, a group of sensor nodes and vehicles are kept underwater. Managing sensor nodes underwater is an expensive technique, and hence, only a few sensor nodes are installed underwater, and to explore and collect data from sensor nodes, vehicles are used [9, 10]. For communication in water, acoustic waves are used, and it has its own challenges to work with like limited bandwidth, underground signal fading, long propagation delay, and high latency. These nodes must have a self-configuring feature. These nodes have limited battery, and due to the ocean environment, battery of nodes cannot be replaced. This technique is very useful for underwater surveillance, pollution monitoring in rivers or oceans, underwater robotics, and most important disaster prevention.

3.4 Underground WSN

In underground WSN, to monitor underground conditions, sensor nodes are placed underground [11]. Additional sink nodes are located aboveground, to convey information from underground sensor nodes to base station [12]. Its maintenance is difficult and costlier than the terrestrial WSN as suitable equipment are required for communication through soil and rocks. Few applications of underground WSN are military border monitoring, landscape management, and agriculture monitoring.

3.5 Terrestrial WSN

In terrestrial WSN, nodes are installed on land in a prearranged area. Its cost is comparatively low, and static routing is used here. In a dense environment, these nodes are capable of efficiently communicating data back to the station [13]. Battery is a major challenge in WSN, and as it has limited power and mostly non-rechargeable, these are used as secondary power source. Applications of terrestrial WSN are surface exploration, industrial monitoring, environmental sensing, etc.

3.6 Application and Challenges of WSN

WSN have been implemented in various application domains. The growth and its use in real world suggest that in the next few decades, sensor nodes will be embedded in most of the objects to make them smart. Its biggest example is the Internet of Things (IoT), and it is making day-to-day items smart by making them communicate with other smart objects and humans. IoT uses sensor nodes for all these works, and the current scenario suggests that it will grow more and will enhance the smartness of objects. Figure 4 depicts the major application areas of WSN.

In military applications, use of WSN helped a lot and still needs further enhancements to improve it. Military applications include security and surveillance, self-healing minefields, border monitoring, search and rescue operations, sniper detection, etc. To detect shooters and to locate them, a counter-sniper named PinPtr

Fig. 4 Major application area of WSNs

is developed [14]. The acoustic shock waves initiated from the gunfire sound, and muzzle blast is sensed by ad hoc acoustic sensor networks. For search and rescue operations, CenWits is used, which determines small area to concentrate for search and rescue operations [15].

Application areas of WSN in agriculture are landscape management, and to save energy in greenhouses, sensors are also used in livestock to keep it healthier, to check soil quality, and to determine micro changes in soil and many more. So, in so many fields, WSN is used, and it is helping farmers and scientists to improve soil quality and to save energy.

WSNs' use in the health care industry is phenomenal in patient monitoring, long-term data collection, care center monitoring, [16] etc. WSN is also used in biodiversity mapping and is very useful for rare flora and fauna species mapping. With the help of sensor nodes, it is easy and effective to find out the locations and count of such rare species and keep track of these.

Natural calamities result in extensive loss of lives and property. That's why disaster management emphasizes on involving multifunctional and multidisciplinary engineering approaches to reduce natural calamities hazards. With the help of WSN, chances of calamities like earthquakes, volcanic eruptions, landslide, floods, storms, blizzards, and droughts can be predicted before time, and preventive actions can be taken accordingly.

WSN's optimization is very much important, and there are many issues and challenges in WSN, which reduce the lifetime and efficiency of WSN. The future research focuses on the different challenges mentioned as below and tries to propose new and innovative solutions to optimize the lifetime and performance of the WSN. Figure 5 depicts the challenges of the WSN.

WSN Challenges Category	Challenges under Category
Design Issues	• Fault • Low latency • Scalability • Transmission media • Coverage problems
Implementation Issues	• Geographic Routing • Sensor Holes • Coverage Topology • Medium Access Schemes • Deployment • Localization • Synchronization

Fig. 5 WSN challenges

4 Meta-Heuristic

Meta-heuristics diverse nature helps in finding the global optima for a problem by skipping the local optima. Algorithms like artificial bee colony, ant colony optimization, and particle swarm optimization have grown a lot in past few years. Each algorithm is based upon different parameters, and they are developed on different characters also, for example, some algorithms are physics inspired, whereas few are bio-inspired. Few of these algorithms are discussed in this section.

4.1 Genetic Algorithm

In 1960, John Holland introduced an algorithm based upon Darwin's theory of evolution. Then after, depending upon different issues, Goldberg and John Holland improved the algorithm. It is used in a number of problems for optimization. The steps of genetic algorithm are as follows [17]:

(i) Create initial population

Repeat

(ii) Provide ranking to the population, according to the fitness
(iii) Cut the weaker solution
(iv) Breed the remaining solutions
(v) Mutate the offsprings

Until termination condition
Terminate

Genetic algorithms' major limitation is the enormous time it takes to find the optimal solution although after enhancements, and it still works better than a few other algorithms.

4.2 Particle Swarm Optimization

James Kennedy and Russell Eberhart proposed population-based optimization algorithm in 1995, on the behavior of swarm particles. The steps of PSO are as follows [18]:

- Initialization
- Initial generation of the particles
- Calculate velocities of the particles
- Update the best solution
- Stopping criteria

In PSO, a group of random particles are assigned, and then by updating generations, optima is searched. Two best solutions are used to update each particle:

- Best solution achieved up to this generation, called pbest.
- Best value achieved by any particle in the population so far. This is the global best called gbest.

Now, the particles update their velocity, using two best values, as shown in Eq. 1

$$\begin{aligned} &\text{v}[\ ]=\text{v}[\ ]+\text{c1}^{*}rand(\ )*(\text{pbest}[\ ]\ \ \text{present}[\ ])+\text{c2}^{*}rand(\ ) \\ &*(\text{gbest}[\ ]\ \ \text{present}[\ ]) \\ &\text{present}[\ ]=\text{present}[\ ]+\text{v}[\ ] \end{aligned} \tag{1}$$

where v[] is the particle velocity, present[] is the current particle (solution), rands() is a random number between 0 and 1, and c1 and c2 are learning factors.

4.3 Ant Colony Optimization

In 1999, Marco Dorigo and Gianni Di Caro proposed a meta-heuristic-based ant colony algorithm and named it as ant colony optimization [19]. It can be applied on discrete optimization problems. It works on the behavior of ants, and ants deposit pheromones on the path that form a trail for other ants. The pheromones on the longer path evaporate and result into the shortest path to be used by default by all other ants. This results in the optimal solution for that problem. This method is used to find the optimal solution for problems like communication, traveling salesman problem, vehicle routing, etc.

4.4 Artificial Bee Colony

Karaboga and Bastruck [21] proposed a population-based algorithm named artificial bee colony. The algorithm execution steps are following:

- Process initialization
- Repeating the process until requirements are not met;

  (a) Employed bees to be kept in the memory for the food sources.
  (b) Onlooker bees also should to be kept in memory on the food sources.
  (c) Scouts will be sent for new food sources searching.

For food source, p_i can be calculated as

$$p_i = \frac{\text{fit}_i}{\sum_{n=1}^{\text{SN}} \text{fit}_n} \tag{2}$$

where solution i's fitness value is symbolized by fit_i and number of food sources is represented by SN. The candidate food position calculation is done by:

$$v_{ij} = x_{ij} + \phi_{ij}\left(x_{ij} - x_{kj}\right) \tag{3}$$

where k 2 1, 2, ..., BN and j 2 1, 2, ..., D are randomly chosen indexes. Eqs. 2 and 3 are used to define the path calculation for food sources.

4.5 Bat Algorithm

Xin-She Yang in 2010 proposed bat algorithm for optimization. It is based on the echolocation behavior of bats [22]. The automatic control and auto-zooming into the region are possible with the help of loudness and pulse emission rates. This acts as the advantage of the algorithm. But this algorithm may lead to stagnation after the initial stage, if the algorithm is allowed to switch to exploitation. Table 1 elaborates the advantages and limitations of the above-explained meta-heuristic algorithms.

Table 1 Advantages and limitation of few meta-heuristic algorithms

Algorithm	Advantages	Limitations	Applications in WSN
Artificial bee colony	Control parameters are less Convergence is fast Both exploitation and exploration	Here, search space is limited by initial solution	Energy efficiency
			Deployment
			Localization
			Routing
			Clustering
Bat algorithm	Automatic control and auto zooming of region	Stagnation after initial stage is very common, if exploitation stage is quick	Deployment
			Clustering
			Localization
Ant colony optimization	It can search among a population in parallel	It has dependent sequences of random decisions. It takes uncertain time for convergence. Here, probability distribution can change for each iteration	Energy efficiency
			Routing
			Clustering
Particle swarm optimization	No overlapping and no mutation calculation. Searching is very fast, with the rate of speed of the particle	For problems of scattering, satisfactory results not found. For non-coordinate systems, technique is not good. Problem of partial optimism	Energy efficiency
			Clustering
			Routing
Genetic algorithm	It can find nearer optimal solutions	It takes a long time to find near optima, but still better than few other algorithms. It is very sensitive to input parameters	Routing
			Energy efficiency

5 Meta-Heuristic Techniques Used in WSN: Literature Review

Meta-heuristic is a good solution to WSN challenges and issues. Some researchers suggest that the optimal solutions provided by the meta-heuristic algorithms are probably the best possible solution for the problem. The solution presented by the different researchers in the literature can be presented below.

Hoang et al. [23] suggest that to extend the lifetime of network operations, optimization is a good solution. HSA protocol has been proposed, and WSNs' lifetime had been extended using this. HSA is more effective in comparison with that of LEACH-C and FCM protocols.

Ado Adamou et al. [24] proposed ABC-SD protocol. ABC optimization algorithms' few fast searching features had been exploited in the proposed algorithm for the clustering process. Kil-Woong Jang [25] proposed an algorithm for minimizing energy consumption in cluster nodes. A meta-heuristic-based channel algorithm is proposed for minimizing energy consumption in WSN. D.C. Hoang et al. [26] proposed the use of their earlier proposed algorithm HSA. According to them, HSA can be used for energy optimization in network and reducing intra-cluster distance. Khalil et al. [27] proposed an algorithm EAERP. It is based on a new evolutionary-based dynamic cluster formation in WSN. The technique guarantees a well-distributed energy consumption.

Khari [28] explained the need for WSN and comparison of different protocols like LEACH, SEP, and TEEN. M. Dhivya et al. [29] focused on data gathering. Cuckoo search optimization technique is applied for data gathering process. Suneet Kumar Gupta et al. [30] focused on the two most important issues of target-based wireless sensor network, coverage, and connectivity. Genetic algorithm is used for solving the above problems, and promising results were found. Stefka Fidanova et al. [32] focused on the issue of the famous telecommunication problem of providing monitoring regions, a full coverage. The motive is to use the tiniest number of sensors and lesser energy consumption for the above problem. Ant colony optimization is used to solve this multi-objective problem. Palvinder Singh Mann et al. [33] proposed an ABC meta-heuristic algorithm. Energy-aware data routing and node clustering in WSN are solved by the proposed algorithm. For efficient clustering of sensor nodes, a cluster head selection algorithm with a multi-objective fitness function is also proposed.

Jain et al. [34] explained different energy aware and shortest path-based route selection approaches to enhance the routing efficiency of the sensor network. Kalpna Guleria et al. [35] focused on the energy-efficient load balanced routing protocols. The authors proposed a novel meta-heuristic ant colony optimization-based unequal clustering (MHACO-UC). The algorithm is useful for optimal path selection and cluster head selection. Lifetime of the cluster head node is increased by the proposed unequal clustering method, rendezvous node (Rnode). Khari et al. [36] explain the recent trends in the field of network security analytics for the perspective of research. Mohsin Masood et al. [37] proposed a meta-heuristic-based algorithm for MPLS/GMPLS networks for selecting efficient routes. BAT-inspired algorithm is used with various levels of loudness parameter, where objective function is routing costs. The aim is to minimize the routing cost. Seyed ali et al. [38] enhanced the capabilities of famous grey wolf optimizer (GWO), to save and retrieve pareto optimal solution. The proposed algorithm is named as multi-objective grey wolf optimizer (MOGWO). This method uses the social hierarchy behavior of grey wolves and their hunting skills in multi-objective space and stimulates them. Selcuk Okdem et al. [39] proposed an application of artificial bee colony algorithm in WSN. The simulation results show that the algorithm provides longer network life and as a result saves more energy.

Muhammad Saleem et al. [40] focused on the principle of swarm intelligence and presented a survey of number of routing protocols based on this technique in WSN. The authors also focused on the applications of swarm intelligence in routing and its general principles. Suneet Kumar Gupta et al. [41] focused on the coverage and connectivity issues of target-based WSN. With the help of genetic algorithm, this NP-complete problem is solved. Chandra Naik et al. [42] focused on extending the lifetime of network in WSN. Differential evolution algorithm that is used to solve target coverage problem is WSN. Oumayma Bahri et al. [43] focused on the multi-objective vehicle routing problem (VRP), with uncertain demands. SPEA2 and NSGAII are two pareto-based evolutionary algorithms, and their extensions are proposed here to solve VRP. Sepehr Ebrahimi Mood et al. [44] proposed a method to determine the best cluster head in each round. The method focused on link quality and energy consumption techniques to find optimal cluster heads. It follows some steps: Firstly calculate optimal number of clusters, then organize the clusters, and then after determine the best cluster head in each round.

Agarwal et al. [45] focused on deep learning models for the detection of DDOS. Deep learning model is used for detection of DDOS on cloud storage. Saini et al. [46] focused on ad hoc network node behavior and different defensive solutions for the same. Vimal et al. [47] explained the use of multi-objective ACO for IoT-based cognitive radio networks. Double Q learning algorithm is used for IoT models with data aggregation and energy-constrained devices.

Tabu search is a meta-heuristic search method based on the next k neighbors' algorithm. Despite the fact that tabu search can handle a wide range of real-world problems, this is the first study of its use to software testing. The structure of TS, which entails searching inside a solution's neighborhood and remembering the best solutions, gives it a simple and obvious technique for creating branch coverage tests. Furthermore, TS is an effective strategy for generating very high branch coverage, based on the experimental findings produced with our tabu algorithm. This paper opens two distinct areas of investigation: the study of algorithm behavior when tabu parameters change and the use of TS to gain various sorts of software coverage [48].

The authors [49] have introduced the G odelTest framework and demonstrated its capabilities on a basic example that is tough for both Boltzmann samplers and Quick Check's analytical and programmatic approaches. The authors also mentioned that in the future, G odelTest might be used to generate test data for more complicated data structures, as well as formulations for local samplers that allow a sufficient degree of freedom to meet bias requirements while staying searchable. They also claimed that G odelTest's separation of the choice model from the program will make it easier to use the same framework for constrained exhaustive generation and machine learning-based dynamic sampler adaption.

The authors [50] created test cases as extended sequences of semantically interacting events using a SA algorithm. Sequences of semantically interacting events were used to create test cases. We also devised an energy function based on the test cases' capacity to encompass a large number of events with a high degree of variation, as well as a definite continuity of events. Our results show that the suggested SO-SA algorithm outperforms the incremental SA method, is competitive with the GA, and outperforms the greedy approach significantly. The SA algorithm will be compared to more meta-heuristics in the future, with the goal of improving the components of the energy function based on larger applications. There are several software test issues that typical software engineering methodologies may not be able to tackle. Nonetheless, similar problems can be theoretically described and solved via mathematical optimization, particularly with the help of metaheuristics.

A new study subject known as search-based software engineering (SBSE) [51] has arisen, which focuses on using optimization approaches to solve software engineering challenges. Because of the importance of the software testing phase, a specific subarea known as search-based software testing (SBST) has grown in importance. This article summarizes the current situation of the area as well as its future possibilities. To begin, we'll go over the most common metaheuristics strategies employed in the field. The state of the art of SBST is then presented through a summary of the primary difficulties that have already been modeled as well as the findings obtained. We can see the potential of this field based on the findings.

Ricca et al. [52] introduced a unified modeling language-based test generating model. These methods go beyond typical path-based testing and incorporate elements of model-based testing. Because the testing models are produced from the web application code, they are categorized as "white box" testing methodologies. The key issues for testing web applications with dynamic features techniques are how to describe the program and what strategy to apply to choose test cases from a large number of options. There hasn't been much study on using state transition diagrams to evaluate web apps with dynamic features.

Marchetto et al. [53] presented a state-based testing technique for Web 2.0 apps to accommodate the new features. The DOM affected by AJAX code is abstracted into a state model in which state transitions are coupled with callback executions triggered by asynchronous messages received from the web server. The test cases are derived from the state model, which is based on the concept of semantically interconnected events. This form of testing has been shown to be effective in detecting flaws via empirical evidence. However, because this technique generates a huge number of test cases, the test suites' use may be limited.

6 Work Done in WSN Using Meta-Heuristic

6.1 Artificial Bee Colony

According to Chandra et al. [54], one of the primary issues in WSN is increasing the life span of a sensor network, which clustering effectively addresses. The usage of the ABC method in cluster head selection is also depicted in multi-hop WSN, which is also documented in this work. The notion of load balancing is used to efficiently utilize energy, and the dynamic channel allocation (DCA) also aids in load balancing inside clustered networks. Not only that, but the numerous simulation results show that DCA may forward packets without causing packet loss. Higher throughput, improved energy efficiency, lower delay, and a longer network lifetime are all advantages of utilizing this technology. The authors concentrated on the WSN's deployment difficulty and used an upgraded version of the ABC algorithm to solve the problem [55].

Ari et al. [24] explained that it's still a problem to build a climbable and energy-efficient sensor network while maximizing its life span. Various laws and regulations have been suggested in order to aid in the management of wireless sensor networks (WSNs). Routing includes significant actions that have a significant impact on the network's quantity and duration. The approach of clustering with data collecting on cluster heads, which also provides efficient salability in such WSNs, significantly improves network life span. This research investigates the ABC algorithm's speedy searching features and competent clustering protocols, as well as a centralized clustering process using data collection techniques and recognition of routing processes in a well-distributed network. The collected results show the efficacy of the suggested protocol in terms of network lifetime and the number of packets transferred.

Because of the large capacity of sensor networks to allow programmers that connect the two worlds, both real and virtual, skilled layout and competent designs of wireless sensor networks have become a hotspot of research. Pawandeep et al. [56] explain that WSN is used in surveillance, medical monitoring, and other applications. They're frequently used to calculate changes in environmental variables such as temperature, pressure, moisture, noise, blood pressure, and heart rate. Sensor nodes are typically made up of a small number of sensors and a mote unit. The sensors in this study are called after the ABC method.

Efficient clustering is a well-known and well-documented optimization topic in wireless sensor networks (WSNs). The artificial bee colony (ABC) metaheuristic is a new addition to a group of computational intelligence approaches that have been applied to WSNs [57]. These techniques include evolutionary algorithms, reinforcement learning, and artificial immune systems. ABC is substantially more common than other population-based metaheuristics in WSNs since it is simple to apply and adaptive. The deficiency in its search equations is due to a poor execution cycle and the demand of specific control parameters. We present an updated version of ABC with a better solution search problem to increase the exploitation capabilities of existing metaheuristics. A better sampling strategy based on the student's t-distribution requires just one control parameter to compute and store, which improves global convergence and hence the efficiency of the suggested metaheuristics. The suggested metaheuristics maintain a good balance between exploration and exploitation with minimum memory requirements, and it also has a first-of-its-kind compact student's t-distribution, making it ideal for the tiny hardware requirements of WSPs. An energy-efficient clustering methodology based on ABC is described in order to make WSNs more energy efficient; it achieves optimal cluster heads along the optical base station location. According to simulation results, the new clustering protocol beats other well-known protocols in terms of packet delivery, energy consumption, network failure, and latency.

Wireless sensor network (WSN) architectures with multiple utility fields, such as army, medical, meteorology, and geology, require reliable communication and powerful routing algorithms [58]. The overall performance of the ABC algorithm on routing tasks in WSNs is investigated in this study. According to the obtained performance results, the protocol in use allows for a longer network life span time by conserving energy. The ABC algorithm is used to assess the complexity of cluster-based routing algorithms. Complete performance and analysis results approve that ABC's set of rules offer promising solutions on WSN routings.

6.2 Ant Colony Optimization

Different problems are optimized using an artificial bee colony-based approach. The following are some problems for which ABC can be utilized as a solution.

Nayyar et al. [59] explained that WSNs encounter a variety of issues and challenges in terms of energy efficiency, limited processing capabilities, routing overhead, packet delivery, and so on. The development of energy-efficient routing protocols has always been a stumbling block for WSNs. To date, a number of routing protocols have been proposed to solve problems based on swarm intelligence. Algorithms based entirely on swarm intelligence are thoroughly investigated and discovered to be adaptable and scalable. This study presents a complete assessment of ant colony optimization-based routing protocols for WSNs in order to give a better platform for academics to work on the many flaws of protocols that have been developed to date.

Fidanova et al. [32] explains that physical or environmental conditions are monitored by WSNs. Full coverage of the tracking location with a minimum number of sensors and network energy consumption is one of the primary goals at some time during their implementation. This problem is difficult, which is why the best way to tackle it is to use some meta-heuristics. This research demonstrates how multi-goal ant colony optimization can be used to overcome this significant telecommunications barrier. Because the number of ANT increases computational time and memory requirements, it is critical to choose the most acceptable range of sellers required to obtain appropriate answers with the least amount of computer resources. As a result, the goal of this paper is to investigate the impact of a wide range of ANTs on the algorithm.

Because WSNs operate on batteries, it is excellent to create routes in them for commercial programmers to perform properly [60]. This work focuses on developing a routing system that enhances a WSN's energy conservation in order to extend battery life. Three parameters have been considered in making a decision for the route to be taken which were the sensor energy in joules, the number of visitors in Erlang, and the space in meters required for a packet to be sent from the supply to the vacation spot node. After that, the routing protocol was integrated into a fuzzy logic and ACO system. The entire node cost to the gateway was calculated using fuzzy logic, which took into account the node's traffic load and energy. The shortest route from the source to the destination sensor node was found and evaluated using ant colony optimization (ACO).

When compared to the overall performance of ACO when observed under similar conditions, the results derived from the MATLAB simulation indicated improved performance in energy conservation.

6.3 Bat Algorithm

The compact bat algorithm [61] is used to solve optimization challenges involving hardware. On probabilistic operations, the proposed algorithm is employed. Nonstop multimodal and WSN uneven clustering are also used to test the technique.

Smart bat algorithm [62] is presented as a way to improve fuzzy approaches and boost searching behavior. The algorithm is implemented in the WSN's 3D environment.

The bat algorithm is used to determine the precision of node localization [31]. The method may also aid with bacterial foraging optimization, according to the researchers. It is particularly successful in resolving WSN-related geographical challenges.

Kavita et al. [63] explain that clustering is a critical topic in WSN. Clusters in the network can be formed based on both the physical location of the nodes and a few overall performance factors. Furthermore, the cluster's nodes are both homogeneous and diverse in character. An aim function is proposed in the suggested technique, and the goal function is used for the implementation of the space among nodes, which is computed using the bat algorithm, and the space is then used for clustering inside the network. The proposed method's impacts are then compared to the core algorithms. The proposed method's impacts are then compared to the core algorithms. The results also validate the algorithm and compare performance factors such as residual energy, end-to-end delay, and network throughput.

Many wireless sensor network (WSN) applications demand information about each sensor node's geographic position. WSN devices are expected to be deployed in huge numbers in a sensing field and self-organize in order to accomplish sensing and acting tasks [64]. The purpose of localization is to assign geographical coordinates to each device in the deployment region that has an unknown position. The application of optimization methods to solve the localization problem has recently become a common strategy. The bat technique is used to estimate the sensor's position in this research.

Because of their diverse uses, wireless sensor networks (WSNs) have received a lot of attention. Node localization is one of the most important challenges in WSN; node localization functionality is reasonably perfect for overall assessment in monitoring systems [65]. Because most sensors cannot recognize their positions due to the value and length of sensors, localization is defined as predicting the locations of sensors from unknown area statistics. The fundamental goal of localization is to find the placements of nodes in a short amount of time with a low energy cost; as a result, new tactics based on swarm intelligence techniques are being used, and node localization is being seen as an optimization problem in a multidimensional space. The meta-heuristic bat algorithm has recently been offered as a solution to the node localization problem. This study introduces Dopeffbat, an efficient solution for the node localization problem that incorporates Doppler effects to improve overall performance. It iteratively computes the nodes' positions using the Euclidian distance (via evolution). When this approach is used on a large WSN with a lot of sensors, it performs well in terms of node localization.

The introduction of modern and various applications in the telecommunications industry has created challenging conditions in the networking field in terms of efficient use of network resources and overall performance optimization [66]. As a result, MPLS/GMPLS (generalized multiprotocol label switching) networks were introduced to provide a better service to fulfil customers' needs while also optimizing network resources. The study employs the bat algorithm with a variety of loudness levels. The simulation findings suggest that MPLS/GMPLS networks of various sizes function better.

Rathour et al. [67] discussed that one of the primary subjects of research in WSN is to compute cluster heads (CH) more precisely and accurately to get better results. WSN is a major field among researches to get better and enhanced network lifetime. Researchers are now focusing on cluster head selection, taking into account a growing number of factors such as residual energy, distance from the base station, and so on. In a recent study, the number of sensors present for a particular sensor node has been used to select a cluster head node. We propose a modified technique that employs the bat algorithm to achieve a CH election that is both optimal and quick.

6.4 Cuckoo Algorithm

Dermi et al. [68] explained that wireless sensor networks (WSN) are a new type of technology that aims to provide modern solutions and capabilities. Their application is growing in a variety of disciplines, including fitness, the environment, and the war. However, the sensor nodes' limited resources are a real barrier, especially in terms of energy autonomy. As a result, one of the most difficult challenges in installing this type of network is developing an energy-conscious routing protocol to extend the system's life span. Clustering is one of the most often used ways for successfully utilizing the network's energy. The latest bio-stimulated optimization method is the Cuckoo search algorithm (CSA). The enhanced Cuckoo search-based (ECSBCP) routing protocol is discussed in this paper as a WSN routing mechanism.

In a sensor network, where each node consumes some amount of power with each transmission over the network, energy competence is the most important attribute. Energy efficiency is required to extend the life of the systems [69]. We shall flow in the course to improve the network life in this document. In addition, on the basis of Cuckoo search, routing can be established to optimize networks in this study. Changes to the PEGASIS (power-efficient gathering in sensor information system) protocol are also made with the help of a fuzzy system. This work is being done to improve network performance as well as network longevity. The goal of the work presented here is to propose an energy-efficient sensor network routing scheme that allows for successful communication without increasing network congestion. Based on the energy analysis, load analysis, and delay analysis, we designed opportunistic routing to discover the most efficient path.

Hosivandi et al. [70] proposed that the limiting of energy sources, which impact the network's lifetime openly, is a significant task in wireless sensor networks. Clustering is a technique for extending the life of a network. Nature-inspired clustering approaches have recently caught the interest of the research community. In this research, we present three Cuckoo algorithm versions in which the power of path length is considered a critical factor in cluster head selection. To avoid cluster heads from quickly dissipating their power, the role of cluster head should be distributed across many nodes. As a result, the suggested methods try to avoid selecting unique nodes as cluster heads as frequently as possible. Furthermore, the problem of not paying attention to the residual energy of sensor nodes during the experimental clustering section of the well-known LEACH algorithm is handled. The proposed algorithms outperform the LEACH algorithm in terms of energy usage and network life span, according to simulation results.

Das et al. [71] explained that the egg-laying radius of the Cuckoo search set of rules is used to construct a cluster on these research paintings, after which the optimum node is sought, which is mostly based on a multi-objective genetic set of rules with pareto ranking, so that the information may be passed to the sink. The overall performance metrics parameters, one of which is the maximizing of network life and the other is the minimization of latency, are the primary focus. When it comes to maximizing the network lifetime parameter, overlapped goal sensing by way of multiple sensors is a waste of energy because the same project may be completed with just one sensor. The series set cover method is used to overcome this problem. The sleep-wake scheduling technique can be used to reduce the delay parameter; however, huge delays are provided since the transmitting node wants to see its next-hop relay node awaken. These delays can be avoided by employing any forged-based packet forwarding mechanism, in which each node sends a packet to the first neighboring node to wake up among a group of prospective nodes. The expected packet-shipping delays from the sensor nodes to the sink node are minimized using this any cast forward technique. The proposed work would provide power-assisted routing with the goal of increasing grid life, packet loss ratio, and system throughput. The suggested technique was compared to the LEACH algorithm in MATLAB. The findings reveal that our suggested approach outperforms the competition in terms of network lifetime, packet loss, and throughput.

Adnan et al. [72] discussed that WSNs are defined by an insufficient amount of electricity delivered to them. As a result, improving an energy green protocol can have a major impact on the community's life span. Verbal exchange is typically the most energy-intensive task that nodes do, and node strength is the principal constraint. A solution could be a good cluster association. Despite the fact that most fulfilling clustering in wireless sensor networks is an NP-hard issue, bio-stimulated meta-heuristic methodologies and strategies are currently extremely popular for solving it. The unusual bio-mimetic Cuckoo search algorithm is used in this paper to provide a centralized power-conscious clustering solution for wireless sensor networks. The cost was calculated with the goal of enhancing network duration while minimizing intra-cluster distance. The proposed algorithm's overall performance is compared to well-known centralized and decentralized clustering protocols. Simulation results show that the offered solutions can extend the lifetime of a network above its competitors.

Chemg et al. [73] explained that the growing popularity of WSN programmers necessitates localization. With the goal of extending the lifetime of the energy-constrained sensor nodes and improving the localization implementation, lowering the computational difficulty, communication overhead in WSN localization is particularly significant. For node localization, this study suggests using the Cuckoo search (CS) algorithm. This strategy, which is mostly focused on changes in step length, allows the population to quickly obtain global top-quality responses, and the fitness of each answer is used to build mutation opportunity for avoiding homegrown convergence. Furthermore, the method limits the population to the defined range in order to reduce energy waste caused by irrelevant searches. Extensive tests were conducted to examine the effects of parameters such as anchor density, node density, and verbal exchange range on the proposed algorithm in terms of average localization mistakes and achievement ratio. In order to determine the ratio, a comparison examination was also performed. When compared to the standard CS algorithm and the particle swarm optimization (PSO) algorithm, experimental results show that the suggested CS algorithm not only boosts convergence rate but also reduces average localization faults.

Kumar et al. [74] discussed that the popularity of wireless sensor networks (WSNs) can be attributed to their ability to manage data with minimal human participation. A large number of sensor nodes are powered by batteries, which are one of the main reasons why energy efficiency is so important in these networks. In light of this, we can observe that the clustering protocol has outperformed the routing protocols and has also shown to be cost-effective. As a result, in comparison to the basic LEACH protocol, this research strongly suggests an innovative clustering technique based on the progeny parasitism of a few cuckoo bird species, which contributes to the increase in the span of these networks. The simulation findings speak for themselves, demonstrating that by incorporating bio-inspired computations into pre-existing protocol designs, network efficiency may be improved significantly.

7 Metaheuristic on WSN Three Major Challenges

As discussed in Sect. 3, there are a number of WSN challenges to work on. This section provides an overview in a tabular manner for three major challenges of WSN, namely, energy efficiency, localization, and routing. Table 2 depicts the short literature survey on energy efficiency problem of WSN. Table 3 depicts the short survey table on localization in WSN using metaheuristic. Table 4 shows how routing efficiency can be improved in WSN using metaheuristic (Table 5).

8 Statistical Analysis

MATLAB 2020 was used, and WSN was created to see how sensor nodes are created, and their linking was analyzed. It was also seen that how all parameters can affect the WSN and how meta-heuristic algorithms can be applied. There are a number of metaheuristic algorithms, but for the analysis purpose, only tabu search algorithm is used. Results are shown in the figures mentioned below. Figure 6 shows the base network. Figures 7 and 8 show the distance to zero and distance to previous nodes, respectively. Figure 9 shows the tabu search implementation on WSN.

Table 2 Literature survey on energy efficiency problem of WSN

S. no.	Authors name	Publication year	Solution provided
[17]	Hoang et al.	2010	Harmony search algorithm is used
[18]	Khalil et al.	2011	EAERP has been proposed with the formulation of the fitness function (ϕEAERP)
[19]	Kil-Woong Jang	2012	Channel scheduling algorithm used for minimizing energy consumption
[20]	Hoang et al.	2014	Harmony search algorithms extension implemented
[21]	Pulido et al.	2014	Multi-objective meta-heuristic
[22]	Jayaprakasam et al.	2014	Meta-heuristic algorithm analysis
[23]	Jayaprakasam et al.	2015	The algorithm named PSOGSA-E had been proposed
[24]	Krishna et al.	2015	MH-EESDA protocol for formation of the secure clusters
[25]	ABBA ARI et al.	2016	Artificial bee colony optimization is used
[26]	Mirjalilia et al.	2016	Multi-objective grey wolf optimization was used to

Table 3 Literature survey on localization problem of WSN

S. no.	Authors name	Publication year	Solution provided
[75]	Hada et al.	2009	Life span method (LSM) is used here
[24]	Jacob et al.	2009	Tabu search and PSO used to improve localization in WSN
[76]	Habib et al.	2010	Coverage restoration scheme is used
[29]	Dhivya et al.	2011	Cuckoo search optimization is applied
[32]	Fidanova et al.	2014	Solution of telecommunication problem using multi-objective ant colony optimization
[31]	Sonia Goyal and Manjeet Singh Patterh	2015	Bat algorithm for node localization
[30]	Gupta et al.	2015	Genetic-based algorithm used to solve the problem
[77]	Arsic et al.	2016	Firework algorithm is applied
[78]	Kaur et al.	2017	The FPA-based node localization algorithm is used

Table 4 Literature survey on routing efficiency of WSN

S. no	Authors name	Publication year	Solution provided
[79]	Srinath et al.	2007	Cluster base routing protocols
[80]	Chen et al.	2009	Inter-cluster forwarding concept
[27]	Enan A. Khalil et al.	2011	Dynamic cluster formation in WSN
[39]	Okdem et al.	2011	Artificial bee colony algorithm is used
[40]	Saleem et al.	2011	Swarm intelligence is used
[81]	Yu et al.	2012	Cluster-based routing for WSNs nonuniform node distribution
[82]	Aslam et al.	2012	Energy-efficient hierarchical routing

Table 5 List of meta-heuristic algorithms applied on these major challenges

WSN issue focused	Meta-heuristic algorithms that were used to optimize the issue
Localisation	Genetic algorithm, scatter search, tab search, differential evolution, ABC, honey bee mating, firefly algorithm, TLBO, ABC, league championship algorithm, gravitational search algorithm, Cuckoo search, Krill herd algorithm, artificial intelligence algorithm, heterogeneous distributed bee algorithm
Routing	Cuckoo algorithm, bat algorithm, killer whale algorithm, hydrological cycle algorithm, mass balance algorithm, Harris hawk optimization, colliding bodies optimization, duelist algorithm, intelligent water drop algorithm, bee's algorithm, glowworm swarm optimization, PSO, ACO
Energy efficiency	Genetic programming, simulated annealing, ACO, reactive search optimization, intelligent water drop algorithm, honey bee mating algorithm, bat algorithm, galaxy-based search algorithm, imperialist based algorithm, rain water algorithm, heterogeneous based bee algorithm, flower pollination algorithm

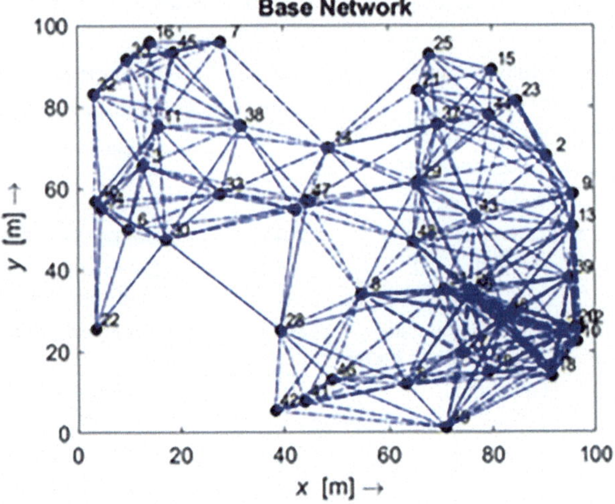

Fig. 6 Base WSN network

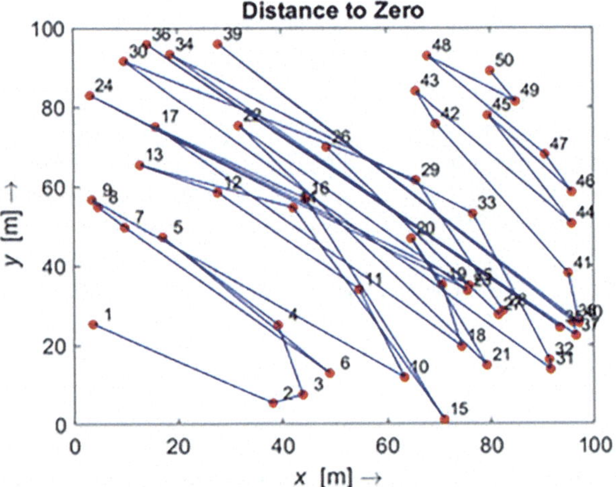

Fig. 7 Distance to zero

9 Conclusion

The survey tries to present the challenges in WSN and the role of meta-heuristic in this field. WSN evolution is rapid, and it is also showing rapid development toward infrastructure deployment and fields like agriculture and health monitoring. A lot of work still can be done in this field for its further enhancement. The fields like IoT and monitoring of remote data cannot be thought of without the WSN. WSN is the base for such fields, and hence, researchers can focus on these fields also by adding features of new upcoming techniques to improve them more. The survey in the paper also shows that the use of meta-heuristic is not limited to WSN and the use of

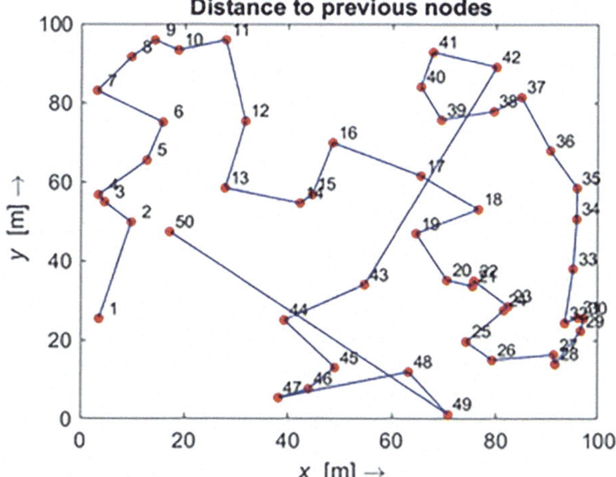

Fig. 8 Distance to previous nodes

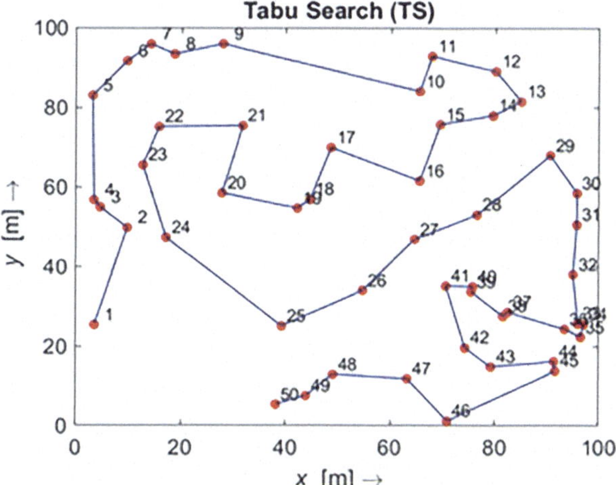

Fig. 9 Tabu search implementation on the implemented WSN for optimization

meta-heuristic in automated software testing is exemplary. In the field of software testing, optimization of test cases and increasing usability are a few tasks which can be optimized with the help of meta-heuristic.

In the future, new algorithms like Newton meta-heuristic algorithms and grey werewolf optimization can be implemented in WSN and automated software testing for optimization.

References

1. Hill, J., Szewczyk, R., Woo, A., Hollar, S., Culler, D., & Pister, K. (2000). *System architecture directions for networked sensors*. ASPLOS.
2. Culler, D. E., & Hong, W. (2004). Wireless sensor networks. *Communication of the ACM, 47*(6), 30–33.
3. Akyildiz, I. F., Su, W. L., Yogesh, S., & Erdal, C. (2002). A survey on sensor networks. *IEEE Communication Magazine, 40*(8), 102–114.
4. Hoang, D. C., Yadav, P., Kumar, R., & Panda, S. K. (2014). Real-time implementation of a harmony search algorithm-based clustering protocol for energy-efficient wireless sensor networks. *IEEE Transactions on Industrial Informatics, 10*(1).
5. Jindal, V. (2018). History and architecture of wireless sensor networks for ubiquitous computing. *International Journal of Advanced Research in Computer Engineering & Technology (IJARCET), 7*(2), ISSN:2278-1323.
6. Yick, J., Mukherjee, B., & Ghosal, D. (2008). Wireless sensor network survey. *Computer Network, 52*(12), 2292–2330.
7. Rawat, P., Singh, K. D., Chaouchi, H., & Bonnin, J. M. (2014). Wireless sensor networks: A survey on recent developments and potential synergies. *Journal of Supercomputing, 68*(1), 1–48.
8. Akyildiz, I. F., Melodia, T., & Chowdhury, K. (2007). A survey on wireless multimedia sensor networks. *Computer Network, 51*(4), 921–960.
9. Akyildiz, I. F., Pompili, D., & Melodia, T. (2004). Challenges for efficient communication in underwater acoustic sensor networks. *ACM SIGBED Review, 1*(2).
10. Heidemann, J., Li, Y., Syed, A., Wills, J., & Ye, W. (2006). Underwater sensor networking: Research challenges and potential applications. *Conference of IEEE Wireless Communications and Networking*.
11. Akyildiz, I. F., & Stuntebeck, E. (2006). Wireless underground sensor networks: research challenges. *Ad Hoc Network, 4*(6), 669–686.
12. Li, M., & Liu, Y. (2007). Underground structure monitoring with wireless sensor networks. In *6th international conference on information processing in sensor networks* (p. 78). ACM.
13. Akyildiz, I. F., Su, W., Sankarasubramaniam, Y., & Cayirci, E. (2002). A survey on sensor networks. *IEEE Communication Magazine, 40*(8), 102–114.
14. Simon, G., Maróti, M., Lédeczi, Á., Balogh, G., Kusy, B., Nádas, A., Pap, G., Sallai, J., & Frampton, K. (2004). Sensor network-based counter sniper system. In *2nd International conference on embedded networked sensor systems* (pp. 1–12). ACM.
15. Huang, J., Amjad, S., & Mishra, S. (2005). CenWits: A sensor-based loosely coupled search and rescue system using witnesses. In *Proceedings of the 3rd international conference on embedded networked sensor systems* (p. 191). ACM.
16. Minaie, A., Sanati-Mehrizy, A., Sanati-Mehrizy, P., & Sanati-Mehrizy, R. (2013). Application of wireless sensor networks in health care system. In *ASES conference and exposition*.
17. Booker, L. B., Goldberg, D. E., & Holland, J. H. (1989). Classifier systems and genetic algorithms. In *Machine learning: Paradigms and methods* (pp. 235–282). MIT Press/Elsevier.
18. Kennedy, & Eberhart, R. C. (1995). Particle swarm optimization. In *Procurement IEEE international conference of neural networks* (Vol. 4, pp. 1942–1948).
19. Dorigo, M., & Caro, G. D. (1999). Ant colony optimization: A new meta-heuristic. In *Proceedings of the congress on evolutionary computation* (pp. 1470–1477).
20. Haldenbilen, S., Ozan, C., & Baskan, O. (2013). *An ant colony optimization algorithm for area traffic control*. INTECH Open Access Publisher.
21. Karaboga, D., & Basturk, B. (2007). An energy efficient routing protocol using ABC to increase survivability of WSN function optimization: Artificial bee colony (ABC) algorithm. *Journal of Global Optimization, 39*, 459–471.
22. Yang, X. S. (2010). A new metaheuristic bat-inspired algorithm. In *Nature inspired cooperative strategies for optimization* (Vol. 284, pp. 65–74). SCI.

23. Hoang, D., Yadav, P., Kumar, R., & Panda, S. (2014). Real-time implementation of a harmony search algorithm-based clustering protocol for energy efficient wireless sensor networks. *IEEE Transaction Industries Informatics, 10*(1), 774–783.
24. Ari, A. A. A., Gueroui, A., Yenke, B. O., & Labraoui, N. (2016). Energy efficient clustering algorithm for Wireless Sensor Networks using the ABC metaheuristic. In *Computer communication and informatics ICCCI international conference on Coimbatore, India.*
25. Jang, K. W. (2012). Meta-heuristic algorithms for channel scheduling problem in wireless sensor networks. *International Journal of Communication Systems, 25*(4), 427–446.
26. Hoang, D. C., Yadav, P., Kumar, R., & Panda, S. K. (2010). A robust harmony search algorithm based clustering protocol for wireless sensor networks. In *IEEE international conference on communications workshops, Singapore* (pp. 1–5).
27. Khalil, E. A., & Attea, B. A. (2011). Energy-aware evolutionary routing protocol for dynamic clustering of wireless sensor networks. *Swarm Evolution Computing, 1*, 195–203.
28. Gopakumar, A., & Jacob, L. (2008). Performance of some metaheuristic algorithms for localization in wireless sensor networks. *International Journal of Network Management, 19*, 355–373.
29. Dhivya, M., & Sundarambal, M. (2011). Cuckoo search for data gathering in wireless sensor networks. *International Journal of Mobile Communication, 9*, 642–656.
30. S. K. Gupta, P. Kuila and P. K. Jana, "Genetic algorithm approach for k -coverage and m -connected node placement in target based wireless sensor networks.", Computers and Electrical Engineering, 2015.
31. Goyal, S., & Patterh, M. S. (2016). Modified bat algorithm for localization of wireless sensor network. *Wireless Personal Communications, 862*, 657–670.
32. Fidanova, S., Marinov, P., & Paparzycki, M. (2014). Multi-objective ACO algorithm for WSN layout: Performance according to number of ants. *International Journal of Metaheuristics, 3*, 149–161.
33. Mann, P. S., & Singh, S. (2016). Artificial bee colony metaheuristic for energy-efficient clustering and routing in wireless sensor networks. *Soft Computing, 21*, 1–14.
34. Mekonnen, M. T., & Rao, N. K. (2017). Cluster optimization based on metaheuristic algorithms in wireless sensor networks. *Wireless Personal Communications, 97*(2), 2633–2647.
35. Guleria, K., & Verma, A. K. (2019). Cluster optimization based on metaheuristic algorithms in wireless sensor networks. *Wireless Personal Communication, 105*(3), 891–911.
36. Arora, S., & Singh, S. (2017). Node localization in wireless sensor networks using butterfly optimization algorithm. *Arabian Journal for Science and Engineering, 42*, 3325–3335.
37. Masood, M., Fouad, M., & Glesk, I. (2017). Proposing bat inspired heuristic algorithm for the optimization of GMPLS networks. In *Proceedings of 25th TELFOR.*
38. Mirjalili, S., Saremi, S., Mirjalili, S. M., & Coelho, L. D. S. (2016). Multi-objective grey wolf optimizer: A novel algorithm for multi-criterion optimization. *Expert System Application, 47*, 106–119.
39. Okdem, S., Karaboga, D., & Ozturk, C. (2011). An application of wireless sensor network routing based on artificial bee colony algorithm. *IEEE Congress of Evolution Computing*, 326–330.
40. Saleem, M., Di Caro, G. A., & Farooq, M. (2011). Swarm intelligence based routing protocol for wireless sensor networks: Survey and future directions. *Information Sciences, 181*(20), 4597–4624.
41. Gupta, S. K., Kuila, P., & Jana, P. K. (2015). Genetic algorithm approach for k -coverage and m -connected node placement in target based wireless sensor networks. *Computation Electrical Engineering.*
42. Naik, C., & Shetty, D. P. (2018). A novel meta-heuristic differential evolution algorithm for optimal target coverage in wireless sensor networks. In *International conference on innovations in bio-inspired computing and applications*. Springer.
43. Bahri, O., Amor, N. B., & Talbi, E.-G. (2018). Possibilistic framework for multi-objective optimization under uncertainty. In *Recent developments in metaheuristics* (pp. 17–42). Springer.

44. Mood, S. E., & Javidi, M. M. (2019). Energy-efficient clustering method for wireless sensor networks using modified gravitational search algorithm. *Evolving Systems*, 1–13.
45. Agarwal, A., Khari, M., & Singh, R. (2021). Detection of DDOS attack using deep learning model in cloud storage application. *Wireless Personal Communications*, 1–21.
46. Saini, R., & Khari, M. (2011). Defining malicious behavior of a node and its defensive techniques in ad hoc networks. *International Journal of Smart Sensors and Ad Hoc Networks, 1*(1), 17–20.
47. Vimal, S., Khari, M., Crespo, R. G., Kalaivani, L., Dey, N., & Kaliappan, M. (2020). Energy enhancement using multiobjective Ant colony optimization with Double Q learning algorithm for IoT based cognitive radio networks. *Computer Communications, 154*, 481–490.
48. Díaz, E., Tuya, J., & Blanco, R. (2003). Automated software testing using a metaheuristic technique based on Tabu search. In *Proceedings of 18th IEEE international conference on automated software engineering* (pp. 310–313). IEEE.
49. Feldt, R., & Poulding, S. (2013). Finding test data with specific properties via metaheuristic search. In *2013 IEEE 24th international symposium on software reliability engineering (ISSRE)* (pp. 350–359). IEEE.
50. Haraty, R. A., Mansour, N., & Zeitunlian, H. (2018). Metaheuristic algorithm for state-based software testing. *Applied Artificial Intelligence, 32*(2), 197–213.
51. de Freitas, F. G., Maia, C. L. B., de Campos, G. A. L., & de Souza, J. T. (2010). Optimization in software testing using metaheuristics. *Revista de Sistemas de Informação da FSMA, 5*, 3–13.
52. Ricca, F., & Tonella, P. (2001). Analysis and testing of web applications. In *Proceedings of the 23rd international conference on software engineering. ICSE* (pp. 25–34). IEEE.
53. Marchetto, A., Tonella, P., & Ricca, F. (2008). State-based testing of Ajax web applications. In *2008 1st international conference on software testing, verification, and validation* (pp. 121–130). IEEE.
54. Soujanya, G. L., & Chandra Mouli, P. V. S. (2017). Energy efficient cluster head selection using ABC with DCA in WSN. *International Journal of Innovative Research in Computer and Communication Engineering, 5*(4).
55. Ajayan, A. R., & Balaji, S. (2013). A modified ABC algorithm & its application to wireless sensor network dynamic deployment. *IOSR Journal of Electronics and Communication Engineering, 4*(6).
56. Pawandeep, M., Garg, M., & Jain, N. (2016). An energy efficient routing protocol using ABC to increase survivability of WSN. *International Journal of Computer Applications (0975 – 8887), 143*(2).
57. Mann, P. S., & Singh, S. (2015). Improved metaheuristic-based energy-efficient clustering protocol with optimal base station location in wireless sensor networks. *Soft Computing*. https://doi.org/10.1007/s00500-017-2815-0
58. Okdem, S., Karaboga, D., & Ozturk, C. (2011). *An application of wireless sensor network routing based on artificial Bee colony algorithm*. 978-1-4244-7835-4/11/$26.00 ©2011. IEEE.
59. Nayyar, A., & Singh, R. (2017). Ant colony optimization (ACO) based routing protocols for wireless sensor networks (WSN): A survey. *International Journal of Advanced Computer Science and Applications (IJACSA), 8*(2).
60. Mualuko, V. M., Kihato, P. K., & Oduol, V. (2017). Routing optimization for wireless sensor networks using fuzzy Ant colony. *International Journal of Applied Engineering Research, 12*(21), 11606–11613. ISSN:0973-4562.
61. Nguyen, T., Pan, J. S., & Dao, T. K. (2019). A compact Bat algorithm for unequal clustering in wireless sensor networks. *Applied Sciences, 9*(1973). https://doi.org/10.3390/app9101973
62. Ng, C. K., Ho Wu, C., Hung Ip, W., & Yung, K. L. (2018). *Smart BAT algorithm for wireless sensor network deployment in 3-D environment, 1089-7798*. IEEE. Personal use is permitted, but republication.
63. Kavita, & Kashyap, R. C. (2016). Improved BAT algorithm based clustering in WSN. *IJEDR, 4*(4), ISSN:2321-9939.
64. Goyal, S., & Patterh, M. S. (2013). Wireless sensor network localization based on BAT algorithm. *International Journal of Emerging Technologies in Computational and Applied Sciences (IJETCAS)*.

65. Mihoubi, M., Rahmoun, A., Lorenz, P., & Lasla, N. (2017). An effective Bat algorithm for node localization in distributed wireless sensor network. *Security and Privacy, 1*, e7. https://doi.org/10.1001/spy2.7
66. Mohsin Masood, S., Fouad, M. M., & Glesk, I. (2017). Proposing Bat inspired heuristic algorithm for the optimization of GMPLS networks. In *25th telecommunications forum TELFOR, Serbia, Belgrade*.
67. Rathour, S. K., & Khan, P. R. (2016). An efficient routing algorithm using Bat algorithm in WSN. *International Journal of Advanced Research in Computer Science and Software Engineering, 6*(12).
68. Dermi, M., Barmati, M.E., & Youcefi, H. (2018). *Enhanced Cuckoo search-based clustering protocol for wireless sensor networks*. 978-1-5386-4238-2/18$31.00. IEEE.
69. Bhatti, G. K., & Raina, J. P. S. (2014). Cuckoo based energy effective routing in wireless sensor network. *International Journal of Computer Science and Communication Engineering, 3*(1).
70. Ghiasiana, A., & Hosivandi, M. (2017). Cuckoo based clustering algorithm for wireless sensor network. *International Journal of Computer (IJC), 27*(1), 146–158.
71. Das, S., Barani, S., Wagh, S., & Sonavane, S. S. (2017). Optimal clustering and routing for wireless sensor network based on cuckoo search. *International Journal of Advanced Smart Sensor Network Systems (IJASSN), 7*(2/3).
72. Md. Akhtaruzzaman Adnan, Razzaque, M. A., Md. Anowarul Abedin, Salim Reza, S. M., & Hussein, M. R. (2016). *A novel Cuckoo search based clustering algorithm for wireless sensor networks*. Springer. Sulaiman, H. A., et al. (Eds.), *Advanced computer and communication engineering technology* (Lecture Notes in Electrical Engineering 362). https://doi.org/10.1007/978-3-319-24584-3_53
73. Cheng, J., & Xia, L. (2016). An effective Cuckoo search algorithm for node localization in wireless sensor network. *Sensors, 16*, 1390.
74. Sandeep Kumar, E., Mohanraj, G. P., & Goudar, R. R. (2014). Clustering approach for wireless sensor networks based on cuckoo search strategy. *International Journal of Advanced Research in Computer and Communication Engineering, 3*(6).
75. Hada, A. K. I. O., & Tsuchiya, R. Y. U. J. I. (2009). A metaheuristic algorithm for wireless sensor network design in railway structures. In *2009 international conference on intelligent sensors, sensor networks and information processing (ISSNIP)* (pp. 231–236). IEEE.
76. Habib, S. J., & Marimuthu, P. N. (2010). A coverage restoration scheme for wireless sensor networks within simulated annealing. In *Seventh international conference on wireless and optical communications networks-(WOCN)* (pp. 1–5). IEEE.
77. Arsic, A., Tuba, M., & Jordanski, M. (2016). Fireworks algorithm applied to wireless sensor networks localization problem. *IEEE Congress on Evolutionary Computation (CEC)*, 4038–4044.
78. Kaur, R., & Arora, S. (2017). Nature inspired range based wireless sensor node localization algorithms. *International Journal of Interactive Multimedia & Artificial Intelligence, 4*(6).
79. Srinath, R., Reddy, A. V., & Srinivasan, R. (2007). Ac: Cluster based secure routing protocol for wsn. In *International conference on networking and services (ICNS'07)* (pp. 45–45). IEEE.
80. Chen, G., Li, C., Ye, M., & Wu, J. (2009). An unequal cluster-based routing protocol in wireless sensor networks. *Wireless Networks, 15*(2), 193–207.
81. Xiu-li, R., Hong-wei, L., & Yu, W. (2008). Multipath routing based on Ant colony system in wireless sensor networks. In *International conference on computer science and software engineering*.
82. Aslam, M., Javaid, N., Rahim, A., Nazir, U., Bibi, A., & Khan, Z. A. (2012, June). Survey of extended LEACH-based clustering routing protocols for wireless sensor networks. In *2012 IEEE 14th international conference on high performance computing and communication & 2012 IEEE 9th international conference on embedded software and systems* (pp. 1232–1238). IEEE.

myCHIP-8 Emulator: An Innovative Software Testing Strategy for Playing Online Games in Many Platforms

Sushree Bibhuprada B. Priyadarshini, Amrut Mahapatra, Sachi Nandan Mohanty, Anish Nayak, Jyoti Prakash Jena, and Saurav Kumar Singh Samanta

1 Introduction to CHIP-8 and Metaheuristics

CHIP-8 is an interpreted programming language. It was developed in the 1970s and became popular since it helped programmers to easily program video games for other machines using metaheuristic optimization solution. The CHIP-8 virtual machine has 4KB (4096 bytes) of memory. The lower 512 bytes of memory, that is, from 0x000 to 0x200, were historically used by the interpreter itself. In our research, where the interpreter is running outside virtual machine's memory space, we will use this area to store the fonts (from 0x00). The programs will be loaded at location 0x200. The machine has 16 8-bit data registers named from V0 to VF out of which 15 (V0 – VE) can be used as general purpose registers (GPRs) for arithmetic and logical operations [1, 2].

We have developed myCHIP-8 emulator employing metaheuristic strategy that moves toward the solution of playing online games through a subset of solutions while exploring upcoming steps [3, 4].The 16th register (VF) serves as an "overflow flag" and is SET when an addition results are in an overflow. In the case of subtraction, the register VF is SET when there is no carry needed. The VF register is also SET when the draw instruction results in a pixel collision. There also exists an index register (I) which is used in the case of memory operations. CHIP-8 instructions or "opcodes" are 16-bits in length [1, 2].

A special register called as program counter (PC) is employed that keeps track of which instruction is to be executed next. Certain instructions like the JUMP

S. B. B. Priyadarshini · A. Mahapatra (✉) · A. Nayak · J. P. Jena · S. K. S. Samanta
Siksha 'O' Anusandhan Deemed to be University, Bhubaneswar, Odisha, India

S. N. Mohanty
College of Engineering, Pune, India

M. Khari et al. (eds.), *Optimization of Automated Software Testing Using Meta-Heuristic Techniques*, EAI/Springer Innovations in Communication and Computing, https://doi.org/10.1007/978-3-031-07297-0_9

instruction (JP addr) can change PC. The stack on the CHIP-8 virtual machine is used to store return addresses at the time when the subroutines are called.

Our implementation is capable of storing up to 16 return addresses. We will maintain a stack pointer (SP) that will always point to the top of the stack. Such a stack pointer is helpful when we want to return from a subroutine. The developed emulator in our approach is basically a metaheuristic optimization approach where high-level procedure is framed to attain good solution toward playing online game [3, 4]. Basically, combinatorial optimization involving various registers is employed in our study.

1.1 Motivation

As architectures grow old, the programs written for them get forgotten. This is mostly because people no longer have access to those older machines. This is somewhat the case for the CHIP-8 architecture too. We wanted to preserve the programs that were written for the CHIP-8 architecture and hence decided to make a CHIP-8 emulator. Other than that, CHIP-8 is also an open-source project which means we could code our instruction and set them into the processor and remove the unnecessary instructions as per our requisites. This has been a huge advantage that open-source architectures are flexible in terms of instruction sets [5].

This is cost-effective as we don't need to be dependent upon specific vendors for their instruction set architectures such as x86, arm Vx, etc., which requires us to have the instructions that we don't need. Modern architectures such as RISC-V are based on principles as like as CHIP-8 in terms of instruction flexibility. CHIP-8 provides an easy way to learn the fundamentals of emulating a CPU. This is both because the instruction set is quite small and the registers are only 8-bits wide.

1.2 CHIP-8 as a Metaheuristic Approach

CHIP-8 is an 8-bit machine that reads the instructions stored in a program into its RAM. It then proceeds to decode the instructions and executes them sequentially with optimality, thereby following a metaheuristic strategy starting from attaining goal through sub-solution design [3, 6, 7]. We have implemented the CHIP-8 machine using C programming language employing metaheuristic strategy. The graphics are handled by using the Simple DirectMedia Layer (SDL2). Our implementation has a "delay timer" which also existed in the original CHIP-8 implementation [3, 4].

When the delay timer is SET, it counts down to zero. CHIP-8 [8] has a 16-key input device (0 – F). Our implementation will map this input configuration to the left side of the keyboard. CHIP-8 has 35 instructions or "opcodes." There has been much debate on whether to employ the "classic switch-case instruction decoding" or "jump tables." Whatever the case, it becomes important to classify the opcodes into separate groups. We can classify the opcodes into five groups according to their first byte:

- Unique: The first byte of these opcodes is unique. For example, 1nnn (JP addr) and 2nnn (CALL addr).
- Begin with "0": The first byte of these opcodes is "0.| For example, 00E0 (CLS) and 00EE (RET).
- Begin with "8": The first byte of these opcodes is "8." For example, 8xy0 (LD Vx, Vy) and 8xy1 (OR Vx, Vy).
- Begin with "E": The first byte of these opcodes is "E." For example, ExA1 (SKNP Vx) and Ex9E (SKP Vx).
- Begin with "F": The first byte of these opcodes is "F." For example, Fx15 (LD DT, Vx) and Fx0A (LD Vx, K).

Our implementation decodes the instructions using function pointers (or jump tables) using optimization strategy. In contrast to the classic switch-case instruction decoding, this simplifies the design since now, we don't have to deal with hundreds of "case" statements. CHIP-8 can display hexadecimal digits (0 – F) as sprites on the 64 × 32 screen [2, 3].

Each sprite is guaranteed to be 5 bytes long. The bits which are SET draw out a pattern that corresponds to the particular character that is to be drawn.

Let us take an example of the character "4."

The binary representation for the character "4" in CHIP-8 is

1001000010010000
11110000
00010000
00010000

The bits which are set (i.e., 1) draw out a pattern similar to that of character "4."

The hexadecimal representation of this binary is as follows:

0x90
0x90
0xF0
0x10
0x10

This is the exact hexadecimal representation that is to be stored in our font array. Since each character is guaranteed to be 5 bytes long and we need to store 16 characters, our font array has to be of 80 bytes long. The font array representation is portrayed in Fig. 1. Thus, the systematic optimization-oriented procedure employed asserts the metaheuristic feature of the proposed emulator [8, 9].

```
uint8_t FONTS[80] = {0xF0, 0x90, 0x90, 0x90, 0xF0, // 0
                     0x20, 0x60, 0x20, 0x20, 0x70, // 1
                     0xF0, 0x10, 0xF0, 0x80, 0xF0, // 2
                     0xF0, 0x10, 0xF0, 0x10, 0xF0, // 3
                     0x90, 0x90, 0xF0, 0x10, 0x10, // 4
                     0xF0, 0x80, 0xF0, 0x10, 0xF0, // 5
                     0xF0, 0x80, 0xF0, 0x90, 0xF0, // 6
                     0xF0, 0x10, 0x20, 0x40, 0x40, // 7
                     0xF0, 0x90, 0xF0, 0x90, 0xF0, // 8
                     0xF0, 0x90, 0xF0, 0x10, 0xF0, // 9
                     0xF0, 0x90, 0xF0, 0x90, 0x90, // A
                     0xE0, 0x90, 0xE0, 0x90, 0xE0, // B
                     0xF0, 0x80, 0x80, 0x80, 0xF0, // C
                     0xE0, 0x90, 0x90, 0x90, 0xE0, // D
                     0xF0, 0x80, 0xF0, 0x80, 0xF0, // E
                     0xF0,0x80,0xF0,0x80,0x80};//F
```

Fig. 1 Font array representation

2 Related Work

The technical reference for the instructions used in our project comes from the work of Thomas P. Greene. Thomas hosts a website. His technical reference to the CHIP-8 instruction set is named "Cowgod's Chip-8 Technical Reference" and is probably the most popular technical reference for the CHIP-8 architecture. The reference has been immensely helpful to us. Although some regard the reference as to be somewhat faulty, it is to be noted that almost all of the available programs for the CHIP-8 architecture are programmed according to this reference. M. Laurence discussed how to write an emulator. Similarly, Jones Nigel (1999) discussed the embedded system programming in details. Likewise, I. Mizutani and J. Mitsugi (2016) discussed a multicode and portable RFID tag events emulator for RFID information system [1, 4].

The implementation idea for the Dxyn (DRW Vx, Vy, nibble) instruction comes from Laurence Muller. Laurence maintains a website. Laurence does an excellent job of explaining the concept of sprites and how the VF register can be used to detect collisions. Further, the website also presented a beautiful method of iterating over a byte bit-by-bit by using a bitwise AND operation [10].

Our implementation also benefited immensely from the advice of not updating the screen after every cycle and instead maintaining a draw flag which makes sure that the screen only gets updated after a draw or a clear screen instruction, that is, the only instructions that are capable of drawing on the screen.

The idea of implementing jump tables instead of huge switch-case statements for decoding opcodes comes from Nigel Jones. The article we referred to was posted on the website. Nigel first posted this article in the May 1999 issue of Embedded Systems Programming. In this article, Nigel goes over the performance benefits from using jump tables over switch-case statements. The article mentions that while the compiler may optimize the switch-case statements into jump tables; however, it is not always the case. It mentions that a programmer cannot reliably predict when a switch-case gets optimized into jump tables [5, 6].

It also provides security methods to apply when implementing jump tables by using the keyword static in our project which would make sure that our table would not be changed by a rogue outsider. This article helped us a lot in implementing jump tables by means of function pointers in our project. It is to be noted that jump tables are preferred over switch-case statements in emulators/interpreters that are far more complex in implementation such as the NES, the SNES, etc. The concept of jump tables proved to be an important lesson for us.

3 Proposed Approach Employing Software Testing (ST)

In this project, we have implemented a CHIP-8 emulator in the C programming language along with all its instructions. As a result, all the games that are available for the CHIP-8 architecture can be run on our emulator. We have developed the emulator while considering all the attributes like *reliability*, *scalability*, *portability*, and *reusability* while playing the online games [11, 18–21].

3.1 mychip-8

The emulator begins by calling the reset() function which resets all the registers (including memory) to zero, seeds the pseudo-random number generator, and loads the fonts at memory location 0x00. It also sets the program counter (PC) to memory location 0x200 at which the to-be-executed program will be loaded. It then proceeds to load the ROM file by calling the loadROM() function. Then it proceeds to create a window via the SDL library. From then, it repeatedly calls the cycle() function until the "Escape" key is pressed, in which case the program exits [2, 12].

The cycle() function is the heart of the emulator and is implemented in the classic "Fetch, Decode and Execute" way. It fetches 2 bytes from the emulator's memory (of the program) and joins them together so as to form an "opcode." Then it increments the PC by 2 since 2 bytes have been fetched from the memory. It masks the opcode with 0xF000 and logically shifts it right by 12-bits so as to get its first byte and consults a jump table for deciding which of the five groups (as discussed earlier) it belongs to.

For example, let us consider the opcode "1208":

Since, we masked 1208 with 0xF000 (and logically shifted it right by 12-bits), we are left with "1." Now, this "1" is looked up on the jump table which decides that the instruction is of the format "1nnn" and calls the SYS_1nnn() function to execute the opcode "1208."

However, we can't do this in case the opcode belongs to the other four groups. This is because, unlike "1208," we cannot immediately decide the exact instruction just by looking at the first byte.

Let us take an example, the opcode "80E0." Looking at its first byte, we can tell that it belongs to the group where opcodes begin with "8." However, we cannot tell whether the instruction belongs to 8xy0, 8xy1 … or 8xy7 just by looking at its first byte. In this case, we also need to look at its last byte.

While implementing the draw instruction, we will wrap around any pixels that are drawn outside the screen. This is in contrast to other available emulators that do not wrap around the pixels. Also, in the implementation of the CXKK instruction that generates a pseudo random number, some games such as VBRIX do not handle the number 32 correctly. Our implementation does not have this limitation (Figs. 2 and 3).

So, our procedure for decoding an opcode becomes as follows:

(i) Get the first byte (first_byte) of the opcode and call JUMP_TABLE[first_byte]

Our jump table is implemented somewhat like this:

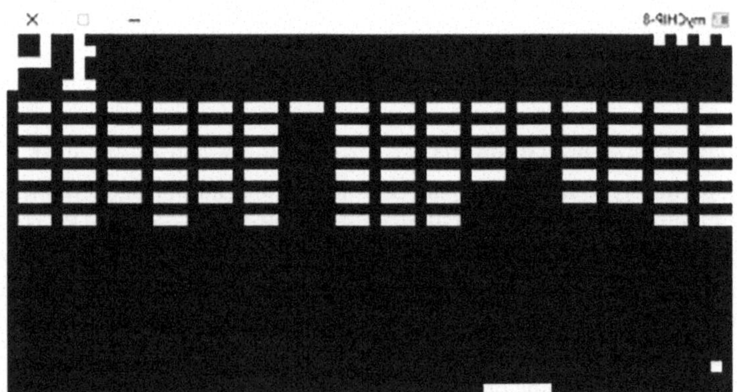

Fig. 2 Entire paddle sprite inside the 64 × 32 display

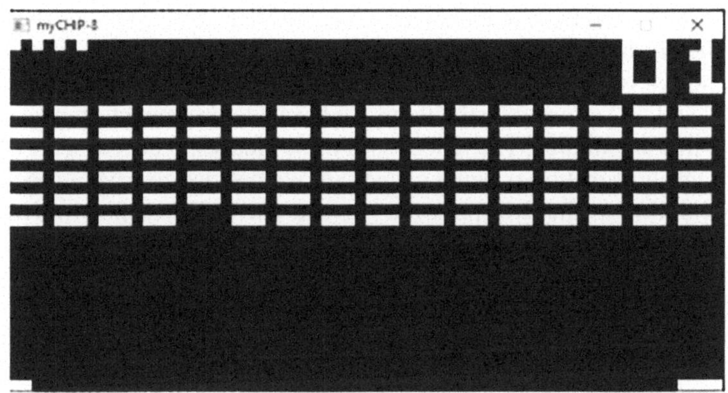

Fig. 3 Part of the paddle that goes outside the screen gets wrap around

```
JUMP_TABLE[first_byte] =


{       JUMP_TABLE_0,           SYS_1nnn,           SYS_2nnn,
SYS_3xkk,
      SYS_4xkk,             SYS_5xy0,              SYS_6xkk,
SYS_7xkk,
        JUMP_TABLE_8,             SYS_9xy0,          SYS_Annn,
SYS_Bnnn,
       SYS_Cxkk,           SYS_Dxyn,             JUMP_TABLE_E,
JUMP_TABLE_F
};
```

(ii) If the first_byte is "unique" (i.e., 1, 2, 3, 4, 5, 6, 7, 9, A, B, C, D), then JUMP_TABLE will call the respective SYS_xxxx() function.
(iii) If the first_byte is "0," then JUMP_TABLE will call JUMP_TABLE_0() function which will further look up the opcode on a jump table which will also look at its last byte to decode it.
(iv) If the first_byte is "8," then JUMP_TABLE will call JUMP_TABLE_8() function which will further look up the opcode on a jump table which will also look at its last byte to decode it.
(v) If the first_byte is "E," then JUMP_TABLE will call JUMP_TABLE_E() function which will further look up the opcode on a jump table which will also look at its last byte to decode it.
(vi) If the first_byte is "F," then JUMP_TABLE will call JUMP_TABLE_F() function which will further look up the opcode on a jump table which will look at its last two bytes to decode it.

The reason for JUMP_TABLE_F() to look at the opcode's last two bytes instead of just one is because unlike others ("0," "8," "E" grouped), the "F" group has two instructions Fx55 and Fx65 that both have their last byte as "5." Obviously, we can't just take the last byte for deciding the exact instruction in this case. Therefore, we consider the last two bytes.

For "E" and "F" grouped instructions, we have a large amount of indexes in our jump table that are unfilled. The functions at these indexes, if called, serve to inform us that an invalid function is being called. So, we decided to have a special function—SYS_INVAL() that will print out "Invalid opcode" and show which opcode failed. It'll then exit [13].

The input handling is done via the SDL2 library which provides a cross-platform graphic library. CHIP-8's inputs and its keyboard mapping are as follows:

```
  KEYPAD        KEYBOARD
 1 2 3 C  =>       1 2 3 4
 4 5 6 D=>       Q W E R
 7 8 9 E      =>       A S D F
 A 0 B F  =>       Z X C V
```

Apart from these keys, the "Esc" (Escape) key can be used at any time to exit the emulator.

4 Instructions Used

(i) SYS_0nnn() – SYS addr

It is ignored by our emulator. This instruction was originally used to call machine language subroutines.

(ii) SYS_00E0() – CLS

This instruction clears the entire 64 × 32 display via a memset() call.

```
memset(VIDEO, 0, sizeof(VIDEO));
```

(iii) SYS_00EE() – RET

This instruction is used to return from subroutines. This is generally used at the end of a subroutine so as to transfer the control back to the position at which the subroutine was called at.

The program counter is set to the address stored at the top of the stack, and 1 is subtracted from the stack pointer.

```
PC = STACK[SP];
SP--;
```

(iv) SYS_1nnn() – JP addr

This instruction jumps to address nnn. This is similar to an unconditional jump. This is usually used to repeat parts of the program one or more times. To get the address, the opcode is masked with 0x0FFF. Then, the program counter is set to that address [6–8].

```
uint16_t address = opcode & 0x0FFFu;
PC = address;
```

(v) SYS_2nnn() – Call addr

This instruction is used to call subroutines which are used to execute parts of the program multiple times. On the COSMAC VIP, since the stack could store up to 12 addresses, 12 subsequent subroutines could be called. In our implementation, 16 subroutines can be subsequently called.

First, the stack pointer is incremented. Then the current value of the program counter (the point at which you're supposed to return after the subroutine ends) is stored onto the top of the stack. To get the address, the opcode is masked with 0x0FFF, and then the program counter is set to the address.

```
SP++;
STACK[SP] = PC;
uint16_t nnn = (OPCODE & 0x0FFFu);
PC = nnn;
```

(vi) SYS_3xkk() – SE Vx, byte

This instruction skips the next instruction (by incrementing the program counter by 2) if the value of register x is equal to kk (a byte). This is an example of conditional branching; however, it is to be noted that only a single instruction can be skipped.

To get the x register, the opcode is masked with 0xF00 and shifted right by 8 bits so as to convert the resultant to a single byte. Then, to get kk (a byte), the opcode is masked with 0x00FF. Finally, we check if the value at register x is equal to that of kk (a byte), and if it is, then skip the next instruction.

```
uint8_t x  = (OPCODE & 0x0F00u) >> 8u;
uint8_t kk = (OPCODE & 0x00FFu);
if(V[x] == kk)
    PC = PC + 2;
```

(vii) SYS_4xkk() – SNE Vx, byte

This instruction skips the next instruction (by incrementing the program counter by 2) if the value of register x is not equal to kk (a byte). This is also an example of conditional branching, but only a single instruction can be skipped.

To get the x register, the opcode is masked with 0xF00 and shifted right by 8 bits so as to convert the resultant to a single byte. Then, to get kk (a byte), the opcode is masked with 0x00FF. Finally, we check if the value at register x is not equal to that of kk (a byte), and if it is not, then skip the next instruction.

```
uint8_t x  = (OPCODE & 0x0F00u) >> 8u;
uint8_t kk = (OPCODE & 0x00FFu);
if(V[x] != kk)
    PC = PC + 2;
```

(viii) SYS_5xy0() – SE Vx, Vy

This instruction skips the next instruction (by incrementing the program counter by 2) if the value of register x is equal to the value of register y. This is also an example of conditional branching, but only a single instruction can be skipped.

To get the first register (x) to be compared, we mask the opcode with 0x0F00 and bitshift it right by 8 bits to convert the resultant into a byte. We mask the opcode with 0x00F0 and bitshift it right by 4 bits to get the second register (y) that is to be compared (x) and convert the resultant to a byte. Then, we compare the values at the two registers (x and y), and if they are equal, we skip the next instruction (by incrementing the program counter by 2).

```
    uint8_t x = (OPCODE & 0x0F00u) >> 8u;
    uint8_t y = (OPCODE & 0x00F0u) >> 4u;

if(V[x] == V[y])
        PC = PC + 2;
```

(ix) SYS_6xkk() – LD Vx, byte

This instruction stores the value of kk (a byte) into the register x.

To get the register x, we mask the opcode with 0x0F00 and bitshift it right by 8 bits so as to convert the resultant into a byte. Then, we get the value of kk (a byte) that is to be stored by masking the opcode with 0x00FF. We then store the value kk in the register x.

```
uint8_t x  = (OPCODE & 0x0F00u) >> 8u;
uint8_t kk = (OPCODE & 0x00FFu);
V[x] = kk;
```

(x) SYS_7xkk() – ADD Vx, byte

This instruction adds kk (a byte) to the already existing byte in the register x and stores it in the register x.

To get the register x, we mask the opcode with 0x0F00 and bitshift it right by 8 bits so as to convert the resultant into a single byte. To get kk (a byte), we mask the opcode with 0x00FF. Then, we add kk (a byte) to the value in the register x and store the result in the register x.

```
uint8_t x  = (OPCODE & 0x0F00u) >> 8u;
uint8_t kk = (OPCODE & 0x00FFu);
V[x] = V[x] + kk;
```

(xi) SYS_8xy0() – LD Vx, Vy

This instruction stores the value in the register y into the register x.

To get the register x, we mask the opcode with 0x0F00 and bitshift it right by 8 bits so as to convert the resultant to a byte. To get the register y, we mask the opcode with 0x00F0 and bitshift it right by 4 bits so as to convert it to a byte. Then we store the value of the register y into the register x [14].

```
uint8_t x = (OPCODE & 0x0F00u) >> 8u;
uint8_t y = (OPCODE & 0x00F0u) >> 4u;
V[x] = V[y];
```

(xii) SYS_8xy1 – OR Vx, Vy

This instruction performs a bitwise OR operation on the contents of the x register and the y register. It then stores the resultant in the x register.

To get the first register (x), we mask the opcode with 0x0F00 and bitshift it right by 8 bits so as to convert the resultant into a byte. To get the y register, we mask the opcode with 0x00F0 and bitshift it right by 4 bits so as to convert the resultant into a byte. Then we perform a bitwise OR operation on the contents of these two registers and store them in the x register.

```
uint8_t x = (OPCODE & 0x0F00u) >> 8u;
uint8_t y = (OPCODE & 0x00F0u) >> 4u;
V[x] = (V[x] | V[y]);
```

(xiii) SYS_8xy2 – AND Vx, Vy

This instruction performs a bitwise AND operation on the contents of the x register and the y register. It then stores the resultant in the x register.

To get the first register (x), we mask the opcode with 0x0F00 and bitshift it right by 8 bits so as to convert the resultant into a byte. To get the y register, we mask the opcode with 0x00F0 and bitshift it right by 4 bits so as to convert the resultant into a byte. Then we perform a bitwise AND operation on the contents of these two registers and store the result in the x register.

```
uint8_t x = (OPCODE & 0x0F00u) >> 8u;
uint8_t y = (OPCODE & 0x00F0u) >> 4u;
V[x] = (V[x] & V[y]);
```

(xiv) SYS_8xy3() – XOR Vx, Vy

This instruction performs a bitwise exclusive OR (XOR) operation on the contents of the x register and the y register. It then stores the resultant in the x register. To get the first register (x), we mask the opcode with 0x0F00 and bitshift it right by 8 bits so as to convert the resultant into a byte. To get the y register, we mask the opcode with 0x00F0 and bitshift it right by 4 bits so as to convert the resultant into a byte. Then we perform a bitwise exclusive on the contents of these two registers and store the result in the x register.

```
uint8_t x = (OPCODE & 0x0F00u) >> 8u;
uint8_t y = (OPCODE & 0x00F0u) >> 4u;
V[x] = (V[x] ^ V[y]);
```

(xv) SYS_8xy4() – ADD Vx, Vy

This instruction adds the content of the x register with the content of the y register and stores them in the register x. If the additive result is greater than 8 bits (i.e., 255), then the carry flag (the VF register) is SET; otherwise it is CLEARED. Only the lowest 8 bits of the additive result is stored in the x register.

To get the x register, we mask the opcode with 0x0F00u and bitshift it right by 8 bits to convert the result into a byte. To get the y register we mask the opcode with 0x00F0 and bitshift it right by 4 bits to convert the result into a byte. We check if the additive result of the contents of the x register and the y register is greater than 8 bits (i.e., 255). The carry flag (VF register) is SET if this condition is true, and it is CLEARED if this condition is false. We then store the additive result in the x register.

```
uint8_t x = (OPCODE & 0x0F00u) >> 8u;
uint8_t y = (OPCODE & 0x00F0u) >> 4u;
if( (V[x] + V[y]) > 255u)
        V[0xF] = 1;
else
        V[0xF] = 0;
V[x] = (V[x] + V[y]);
```

(xvi) SYS_8xy5() – SUB Vx, Vy

This instruction checks if the content of the x register is greater than the content of the y register. If the condition is true, then the carry flag (VF register) is SET; otherwise, it is CLEARED.

To get the x register, we mask the opcode with 0x0F00 and bitshift it by 8 bits so as to convert the resultant into a byte. To get the y register, we mask the opcode with 0x00F0 and bitshift it right by 4 bits so as to convert the resultant into a byte. We then check if the content of the x register is greater than the y register. If the condition is true, we SET the carry flag (VF register); otherwise it is CLEARED. We then subtract the content of the y register from the content of the x register (i.e., V[x] – V[y]) and store the result in the x register.

```
uint8_t x = (OPCODE & 0x0F00u) >> 8u;
 uint8_t y = (OPCODE & 0x00F0u) >> 4u;
 if(V[x] > V[y])
         V[0xF] = 1;
  else
         V[0xF] = 0;
  V[x] = (V[x] - V[y]);
```

(xvii) SYS_8xy6() – SHR Vx {, 2Vy}

This instruction checks if the least significant bit of the content of the x register is 1. If the condition is true, VF register is SET; otherwise it is cleared.

To get the x register, we mask the opcode with 0x0F00 and bitshift it right by 8 bits so as to convert the resultant to a single byte. We then perform a bitwise AND operation on the content of the x register and the number 0x1 and store the result in the VF register (this checks if the least significant bit is 1 or not). We then bitshift the content of the x register by 1 bit so as to divide it by 2.

```
uint8_t x = (OPCODE & 0x0F00u) >> 8u;
V[0xF] = (V[x] & 0x1u);
V[x] = (V[x] >> 1);
```

(xviii) SYS_8xy7 – SUBN Vx, Vy

This instruction subtracts the content of the register x from that of y and checks if the value in y register is greater than that of x. If this condition is true, then the VF register is SET; otherwise, it is CLEARED.

To get the x register, we mask the opcode with 0x0F00 and bitshift it right by 8 bits. To get the y register, we mask the opcode with 0x00F0 and bitshift it right by 4 bits. We then check if the content of the y register is greater than that of the x register. If the condition is true, we SET the VF register and CLEAR the VF register if it is not. We then subtract the content of the y register from the content of the x register and store the resultant in the x register.

```
uint8_t x = (OPCODE & 0x0F00u) >> 8u;
 uint8_t y = (OPCODE & 0x00F0u) >> 4u;
 if(V[y] > V[x])
         V[0xF] = 1;
 else
         V[0xF] = 0;
 V[x] = (V[y] - V[x]);
```

(xix) 8xyE – SHL Vx, {, Vy

This instruction checks if the most significant bit of the content of the x register is 1. If the condition is true, then the VF register is SET; otherwise, it is CLEARED. Then, the content of the x register is multiplied by 2. To get the x register, we mask the opcode with 0x0F00 and bitshift it right by 8 bits so as to convert it into a single byte. We then bitshift the content of the x register by 7 bits right (to get the most significant bit) and store it in the VF register. This makes sure that the VF register is SET if the most significant bit is 1 (i.e., the value is greater than 128 since it is the only way for the MSB to be 1 for a 8 bit value) and is CLEARED otherwise. We then bitshift the content of the x register left by 1 bit so as to multiply it by 2 [7–10, 15].

```
uint8_t x = (OPCODE & 0x0F00u) >> 8u;
V[0xF] = V[x] >> 7u;
V[x] = (V[x]  << 1);
```

(xx) 9xy0 – SNE Vx, Vy

This instruction checks if the content of the x register is not equal to that of the y register. If the condition is true, then the program counter is incremented by 2 (effectively skipping the next instruction).

To get the x register, we mask the opcode with 0x0F00 and bitshift it right by 8 bits so as to convert it into a single byte. To get the y register, we mask the opcode with 0x00F0 and bitshift it right by 4 bits so as to convert it into a single byte. We then check if the content of the x register is not equal to the content of the y register. If the condition is true, then the program counter is incremented by 2.

```
uint8_t x = (OPCODE & 0x0F00u) >> 8u;
uint8_t y = (OPCODE & 0x00F0u) >> 4u;
if(V[x] != V[y])
        PC = PC+2;
```

(xxi) Annn – LD I, addr

This instruction sets the value of the register I to nnn (an address).

To get the address we mask the opcode with 0x0FFF and store it in the variable named nnn. We then set the I register to the variable nnn.

```
uint16_t nnn = (OPCODE & 0x0FFFu);
I = nnn;
```

(xxii) Bnnn – JP V0, addr

This instruction sets the program counter to V0 + addr (i.e., the content of the 0th register + the address nnn) effectively jumping to that address. This is also an example of an unconditional jump.

To get the address (that is to be added to the content of the 0th register), we mask the opcode with 0x0FFF and store it in the variable named nnn. We then set the program counter to the additive result of that of the content of the 0th register and the address nnn.

```
uint16_t nnn = (OPCODE & 0x0FFFu);
PC = (V[0] + nnn);
```

(xxiii) Cxkk – RND Vx, byte

This instruction performs a bitwise AND operation between the contents of the x register and kk (a byte). It then stores the result in the x register.

To find out the x register, we mask the opcode with 0x0F00 and bitshift it right by 8 bits. To find kk (the byte), we mask the opcode with 0x00FF. We then perform a bitwise and between a random number (that is generated by our PRNG—pseudo random number generator) and kk (the byte). The result of this operation is stored in the x register [16, 17].

```
uint8_t x  = (OPCODE & 0x0F00u) >> 8u;
uint8_t kk = (OPCODE & 0x00FFu);
uint8_t prng_b = rand() & kk;
V[x] = prng_b;
```

(xxiv) Dxyn – DRW Vx, Vy, nibble

This instruction displays (on the 64 × 32 screen) a sprite that starts at the memory location pointed to by the I register. The sprite is displayed at the position pointed by the x and the y register. Sprites are XORed onto the screen, and if this causes any of the previous sprites on the screen to be turned off, then the VF register is SET.

This is the only instruction that's capable of drawing on the screen and hence the most important instruction of this whole project. To get the x register, we mask the opcode with 0x0F00 and bitshift it right by 8 bits so as to convert the resultant into a byte. To get the y register, we mask the opcode with 0x00F0 and bitshift it right by 4 bits so as to convert the resultant into a byte. The value "n" represents the height of the sprite. To get n we mask the opcode with 0x000F. We then declare two variables xc and yc.

The content of the x register is stored in xc, and the content of the y register is stored in yc. This pair xc and yc denotes the position at which the sprite is to be drawn. In our first loop (named outside_loop), we iterate from "0" to "n" which is the number of rows to be drawn. With each iteration, we load the sprite byte from memory[I + outside_loop_counter] into the variable sprite_byte. For each such row, we also need to iterate over the byte, bit by bit. This becomes our second loop (named inside_loop). Since a sprite is 8 bits long, we need eight iterations to iterate over an entire row. So, we iterate from "0" to "8" for each bit and check if the bit is SET (i.e., 1). If it is SET, then we need to draw a pixel at the position yyPos=(yc + outside_loop_counter) and xxPos=(xc + inside_loop_counter). yyPos is named so because it represents the y-coordinate at which the pixel of the sprite is going to get drawn. The same goes for xxPos. Since pixels are XORed onto the screen, we must check for any existing pixel at the point where we are about to draw (because for an XOR operation, the result is 0 if both the operands are 1). If there is such an existing pixel, then we SET the VF register. Here, the VF register acts as a collision detector. Finally, we draw at the position (xxPos, yyPos).

```
        uint8_t x = (OPCODE & 0x0F00u) >> 8u;
        uint8_t y = (OPCODE & 0x00F0u) >> 4u;
        uint8_t n = (OPCODE & 0x000Fu);
        uint8_t xc = V[x];
        uint8_t yc = V[y];
        V[0xF] = 0;
for(uint8_t i=0 ; i<n ; i++)
        {
                uint8_t sprite_byte = MEMORY[I + i];
for(uint8_t col=0 ; col<8 ; col++)
                {
                        uint8_t sprite_bit = sprite_byte& (0x80u
>> col);
                        uint8_t yyPos = yc + i;
                        uint8_t xxPos = xc + col;
xxPos = xxPos%64;
yyPos = yyPos%32;
        uint32_t *spp = &VIDEO[(yyPos) * 64 + (xxPos)];
                        if(sprite_bit)
                        {
```

```
if(*spp == 0xFFFFFFFF)
                                        V[0xF] = 1;
                              *spp ^= 0xFFFFFFFF;
                    }
          }
}
Notice how be iterate over an entire byte, bit-by-bit:
for(uint8_t col=0 ; col<8 ; col++)
          {
                    uint8_t sprite_bit = sprite_byte& (0x80u
>> col);

Let's say our sprite_byte is 01011010


When col = 0, sprite_bit = 01011010 & (0x80u >> 0)
i.e.
    0 1 0 1 1 0 1 0
&    1 0 0 0 0 0 0 0     (0x80u >> 0 = 10000000 >> 0 = 10000000)
    --------------------
    0 0 0 0 0 0 0 0
    --------------------
Similary,
When col = 4, sprite_bit = 01011010 & (0x80 >> 4)
i.e.
    0 1 0 1 1 0 1 0
    0 0 0 0 1 0 0 0     (0x80u >> 4 = 10000000 >> 4 = 00001000)
&    --------------------
    0 0 0 0 1 0 0 0
    --------------------
```

(xxv) Ex9E – SKP Vx

This instruction increments the program country by 2 if the key with the value represented by the content of the x register is pressed.

To get the x register, we bitmask the opcode with 0x0F00 and bitshift it right by 8 bits so as to convert the resultant into a byte. Then we check if the key with the value of the x register is pressed. If the condition is true, then the program counter is incremented by 2 (effectively skipping the next instruction).

```
      uint8_t x = (OPCODE & 0x0F00u) >> 8u;
        uint8_t k = V[x];
if(KEYPAD[k])
                    PC = PC + 2;
```

(xxvi) ExA1 – SKNP Vx

This instruction increments the program country by 2 if the key with the value represented by the content of the x register is not pressed.

To get the x register, we bitmask the opcode with 0x0F00 and bitshift it right by 8 bits so as to convert the resultant into a byte. Then we check if the key with the value of the x register is not pressed. If the condition is true, then the program counter is incremented by 2 (effectively skipping the next instruction).

```
uint8_t x = (OPCODE & 0x0F00u) >> 8u;
uint8_t k = V[x];
if(!KEYPAD[k])
        PC = PC + 2;
```

(xxvii) Fx07 – LD Vx, DT

This instruction loads the x register with the value of the delay timer.

To get the x register, we mask the opcode with 0x0F00 and bitshift it right by 8 bits so as to convert the resultant into a byte. Then we place the value of the delay timer register into the register x.

```
uint8_t x = (OPCODE & 0x0F00u) >> 8u;
   V[x] = DT;
```

(xxviii) Fx0A – LD Vx, K

This instruction waits for a key press. Until a key is pressed, all execution stops.

To get the x register, we mask the opcode with 0x0F00 and bitshift it right by 8 bits so as to convert the resultant into a byte. We then check if any of the 16 keys is pressed. If the condition is true, then the normal execution continues. If this condition is false, then we decrement the program counter by 2. This effectively means that until a key is pressed our program will not continue forward.

```
        uint8_t x = (OPCODE & 0x0F00u) >> 8u;
        bool flag = false;
for(int k=0 ; k<16; k++)
if(KEYPAD[k])
                {
                        V[x] = k;
                        flag = true;
                        break;
                }
if(flag == false)
                PC = PC - 2;
```

(xxix) Fx15 – LD DT, Vx

This instruction loads the delay timer register with value present in the x register.

To get the x register, we mask the opcode with 0x0F00 and bitshift it right by 8 bits so as to convert the resultant into a byte. We then get the value present in the x register and load it into the delay timer register.

```
uint8_t x = (OPCODE & 0x0F00u) >> 8u;
DT = V[x];
```

(xxx) Fx18 – LD ST, Vx

This instruction loads the sound timer register with the value present in the x register.

This instruction is ignored by our interpreter.

(xxxi) Fx1E – ADD I, Vx

This instruction adds the value of the I register and the value of the x register. It then places the result into the I register.

To get the x register, we mask the opcode with 0x0F00 and bitshift it right by 8 bits so as to convert the resultant into a byte. We then add the value in the I register and the value present in the x register. This additive result is placed in the I register.

```
uint8_t x = (OPCODE & 0x0F00u) >> 8u;
I = I + V[x]
```

(xxxii) Fx29 – LD F, Vx

This instruction places the memory address of the sprite of the digit (present in the x register) into the I register.

To get the x register, we mask the opcode with 0x0F00 and bitshift it right by 8 bits so as to convert the resultant into a byte. We then find the digit (0x0 – 0x16 i.e., 0 – F) of the sprite whose position is to be loaded in the I register. We then multiply the digit by 5 (to get the offset at which the said digit resides in memory) and place this result in the I register.

```
uint8_t x = (OPCODE & 0x0F00u) >> 8u;
uint8_t digit = V[x];
I = (5 * digit);
```

(xxxiii) Fx33 – LD B, Vx

This instruction stores the BCD (binary coded decimal) value of the content of the x register into the memory locations I, I + 1 and I + 2.

To get the x register, we mask the opcode with 0x0F00 and bitshift it right by 8 bits so as to convert the resultant into a single byte. Since an unsigned 8-bit number can only go up to 255, we find out the last digit and place it into MEMORY[I + 2].

Then we find out the second-last digit and place it into MEMORY[I + 1]. And finally we find out the third last digit, that is, the first digit, and place it into MEMORY[I]. This is done by a series of repeated divisions and modulo operations.

```
        uint8_t x = (OPCODE & 0x0F00u) >> 8u;
        uint8_t number = V[x];
MEMORY[I + 2] = number % 10;
        number = number / 10;
MEMORY[I + 1] = number % 10;
        number = number / 10;
MEMORY[I + 0] = number % 10;
```

(xxxiv) Fx55 – LD [I], Vx

This instruction places the contents of the registers beginning from the 0 register to the x register into the memory location starting at the value of the I register.

To get the x register, we mask the opcode with 0x0F00 and bitshift it right by 8 bits so as to convert the resultant into a byte. We then place the contents of the registers beginning from the 0 register to the x register starting at memory location I.

```
        uint8_t x = (OPCODE & 0x0F00u) >> 8u;
for(uint8_t r=0 ; r<=x ; r++)
MEMORY[I + r] = V[r]
```

(xxxv) Fx65 – LD Vx, [I]

This instruction places the contents of the memory location starting at the value of the I register into the registers beginning from the 0 register to the x register.

To get the x register, we mask the opcode with 0x0F00 and bitshift it right by 8 bits so as to convert the resultant into a byte. We then place the contents of the memory location starting at the value of the I register into the registers beginning from the 0 register to the x register.

```
        uint8_t x = (OPCODE & 0x0F00u) >> 8u;
for(uint8_t r=0 ; r<=x ; r++)
                V[r] = MEMORY[I + r]
```

5 Result Discussion and Debugging Issues

As soon as the emulator was finished, it was clear to us that the code would require some debugging to remove bugs for automated software testing [9]. Some games like to invoke a particular instruction that was present on the real machine on which

CHIP-8 was originally programmed and was not available in our machines. The emulator was then strictly directed to ignore such instruction. It would process the instruction, but the instruction itself will not do anything. Another problem was of pixel wrapping. Some games allowed pixels to go out of the screen. This was a problem for us since the display was of fixed size and accessing anything out of the video buffer resulted in a segmentation fault. This was fixed by making sure that pixels that would be drawn outside the screen will now be wrapped around the screen. We also fixed two logical instructions that behaved a bit differently than what our technical reference claimed.

The problem was that most references directed to check for strictly greater than or strictly less than comparisons, but they never cited what would happen in case of equality. If we ignored equality cases, the two instructions would work improperly, and in a particular game, they would result in a sprite getting redrawn when it was not supposed to. The unit and integration testing employed in our approach ensure the reliability and portability of our proffered system.

Debugging this issue was particularly tedious because it required to step over instructions of the said game one by one to see which particular instruction was causing the issue. It was also tedious because of the fact that we did not expect instructions that did logical comparisons to be responsible for the sprite redrawing bug in the said game. As a result, we spent days going over our draw instruction and making sure it was correct before realizing that it was the comparison instruction that was the problem. With these bugs ruled out, our emulator worked perfectly with all games that we tested on it.

6 Conclusions

Our primary motivation behind choosing to build a basic emulator was to understand the fundamentals of how CPUs worked. We were always fascinated by microprocessors especially the older ones such as the 6502, Z80, 8086, m68k, etc. These were very successful CPUs that were used in many video game consoles while considering testing. For example, the 6502 was used on the NES and the Z80 on the Pac-Man arcade machine. Also, the legendary personal computer, the Commodore Amiga, was powered by one of these processors as well—the m68k. This project gave us an insight into writing emulators while considering both reusability, portability, and reliability of the software developed, and we hope the knowledge gained from this project will help us write more complex emulators in the future.

Acknowledgment We would like to thank the Department of Computer Science & Information Technology, Institute of Technical Education & Research, Siksha 'O' Anusandhan Deemed to be University for making this investigation successful.

References

1. http://www.multigesture.net/articles/how-to-write-an-emulator-chip-8-interpreter/
2. http://devernay.free.fr/hacks/chip8/C8TECH10.HTM
3. Osman, I., & Kelly, J. (1996). *Meta-hueuristics theory and applications* (pp. 1–690) ISBN:978-1-4613-1361-8.
4. Sharma, M., & Kaur, P. (2021). A Comprehensive analysis of nature-inspired meta-heuristic techniques for feature selection problem. *Archives of Computational Methods in Engineering, 28*, 1103–1127.
5. Jones, N. (1999). Arrays of pointers to functions: Embedded systems programming. *Embeded Systems Programming*, 46–56.
6. Martinez, F., Herrero, C., & Pablo, S. D. (2014). Open loop wind turbine emulator. In *Renewable energy* (Vol. 63, pp. 212–221). Elsevier.
7. Mizutani, I., & Mitsugi, J. (2016). A multicode and portable RFID tag events emulator for RFID information system. In *Proceeding of 6th international conference on internet of things* (pp. 187–188).
8. Babacan, Y., & Kakar, F. (2016). *Floating memristror emulator with subthreshold region* (Vol. 90, pp. 471–475). Springer.
9. Priyadarshini, S. B. B., & Panigrahi, S. (2017). A distributed scalar controller selection schemefor redundant data elimination in sensor networks. *International Journal of Knowledgediscovery in Bioinformatics, IGI Global, 7*(1), 91–104.
10. Ahrenholz, J., Goff, T., & Adamson, B. (2011). Integration of the CORE and EMANE network emulators. In *MILCOM 2011 military communications conference*.
11. Jena, A., Das, A. K., Mohapatra, H., & Prasad, D. (2020). *Automated software testing foundations, applications and challenges* (pp. 1–165) ISBN:978-981-15-2455-4.
12. Ellims, M., & Jackson, K. *ISO 9001: Making the right mistakes* (SAE Technical Paper Series 2000-01-0714).
13. Davis, M., & Weyuker, E. J. (1998). Metric-based Test data Adequacy criteria. *Computer Journal, 13*(1), 17–24.
14. Jorgensen. (1995). *Software testing a craftsman's approach*. CRC Press.
15. Lyu, M. R., Huang, Z., Sze, S. K., & Cai, X. (2003). *An empirical study on testing and fault tolerance for software reliability engineering* (pp. 119–130).
16. McCabe, T. J. (1976). A complexity measure. *IEEE Transactions on Software Engineering, 2*(4), 202–213.
17. Richard, D. J., & Thompson, M. C. (1993). An analysis of test data selection criteria using the relay model of fault detection. *IEEE Transactions on Software Engineering, 19*(6), 533–553.
18. Torkar, R., Mankefors, S., Hansson, K., & Jonsson, A. (2003). An exploratory study of component reliability using unit testing. In *Proceeding 14 international symposium on system reliability engineering* (pp. 227–233).
19. Voas, J. M., & Miller, K. W. (1995). Software testability: The new verification. *Software*, 17–18.
20. Zhu, H. (1996). A formal analysis of subsume relation between software testing adequacy criteria. *IEEE Transactions on Software Engineering, 22*(4), 248–255.
21. Zhu, H., Hall, P. A. V., & May, J. H. R. (1997). Software unit test coverage and adequacy. *ACM Computing Surveys, 29*(4), 366–427.

Defects Maintainability Prediction of the Software

Kanta Prasad Sharma, Vinesh Kumar, and Dac-Nhuong Le

1 Introduction

Software development does not end with the build of software, as there is an extreme need to maintain the software even after it is deployed or built, and this leads to the advent for research in the software maintenance field to find the effective method to predict the software maintainability. The identification and bug removable from the real-time working software is called software maintainability which allocates specific resources to the software to keep it working on. Various researches have been performed to solve this issue that aims to build a model that can predict the software maintainability efficiently so that the bugs can be fixed and new functionalities can be added to the software [1]. This resolution of software issues aids the organizations to have a very high impact on the industry. As software maintainability is enhanced by fixing bugs and removing defects, the defects should also have an impact on tracking and finalizing the software maintainability [2]. Hence, this research provides a proposed method based on the implementation of two experiments for the impact of defects on software maintainability detection problem by calculating the MI score that classifies the problem into two classes on the basis of a threshold or fixed value [3–5].

In the modern era, the role of software maintainability has become a fundamental requirement for existing as well as upcoming software. The latest techniques and autonomous software measure performance and support for decision-making in the industry [6]. The proposed research work provides a review status on the previous

K. P. Sharma · V. Kumar
University Institute of Computing, Chandigarh University Mohali, Ludhiana, Punjab, India
e-mail: vinesh.e11095@cumail.in

D.-N. Le (✉)
Faculty of Information Technology, Haiphong University, Haiphong, Vietnam
e-mail: nhuongld@dhhp.edu.vn

M. Khari et al. (eds.), *Optimization of Automated Software Testing Using Meta-Heuristic Techniques*, EAI/Springer Innovations in Communication and Computing, https://doi.org/10.1007/978-3-031-07297-0_10

factor of software maintainability assessment based on five existing developed models. These models are imposed in the industrial software systems for auto mode performance analysis, and some models are also useful for their effectively involvement as industrial software systems [7].

The main motive of the proposed work is to show the impact of defects on maintainability prediction. The flow of paper is managed in six sections. The overview based on a literature review on the software maintainability test in Sect. 2. Section 3 describes the fundamental knowledge that is needed in order to perform the proposed work. Section 4 describes the methodology that is designed to deploy the proposed work. In Sect. 5, experimental results are presented that satisfactorily prove the aim of the proposed work. Finally, Sect. 6 concludes the paper presenting the reasons for the efficacy of the proposed work.

2 Literature Survey

This section describes some of the studies that have been carried out in the respective field.

Propounded different features are directly affected on the maintainability of the software such as testability, stability, changeability, and analyzability [8]. They also analyzed the effect of various complexity measures like `RFC, WMC, DIT, CBO`, and `NOC` on the stated factors/features. The improvements on complexity measures as `CBO_U, CBU_IUB, CBO_NA, NOC` and also explored the sub-features of software with inheritance, linking of modules, space and time performance analysis, and that cause of maintainability prediction of the working software [1, 9].

Ghosh et al. proposed a three-feature model in which they considered various components that affect the complexity, such as nested components, operations, and attributes [2, 10]. The issue of software maintainability is used the vulnerability knowledge of software is predicted using a self-learning approach. SMP learner is to aid learning of maintainability prediction models using self-learning approach by employing 44 four-level code measures. In addition, they utilized 150 experimental results to evaluate the SMP learner [13, 14]. In this chapter, the proposed model outperform indicate as the behavior four metrics based MI model as well as other traditional maintainability prediction models.

The SMEM-MCC model is used to evaluate software maintainability utilizing combination of classifiers [4, 15]. Firstly, they used generic algorithm to select the attributes that are most relevant for software maintainability prediction. Thereafter, they classified 300 software class metrics using decision tree classifier and obtained outperforming results than SMO, BPNN, and simple classifier. Propounded enhanced SVM-based technique [5], SVMSBCTC to detect identical bugs' report [16]. They considered three features based on textual correlation and three OS projects—SVN, ArgoUML, and Apache—to execute the proposed research work. Their experimental results have shown the proposed SVMSBCTC as an outperformer when compared with the SVM-54 method [17, 18].

Deep learning has also been taken into consideration for software maintainability metrics prediction since the last few years. A deep learning principal is used for real-time prediction and maintainability on 299 software data using five machine learning models [6]. They have exploited 29 OO metrics to accomplish the proposed work. The outcomes have shown that their proposed LSTM model outperformed other machine learning baselines taken into consideration [19].

MI metric given by Oman and Hagemeister [7] is widely used to predict software maintainability since the last several studies. Maintainability of conventional code blocks can be easily detected using MI metric, and the object-oriented code blocks can be easily taken into consideration for maintainability check using this metric after its adaptation by OO with some specific limitations. Welker [8] check the validity of MI metric on software maintainability prediction. The experimental results shown by him also gave evidence for the validity of MI measure on conventional code blocks. Aggrawal et al. [9] have propounded ANN principal adopted to enhance the maintainability performance of object-oriented code utilizing OO metrics.

Thwin et al. [10] have also proposed NNets to estimate the quality of software utilizing the OO metrics. Zhou et al. [11] have propounded software applications to find metrics used to detect the maintainability of software. Zhou et al. [12] provide the connection between the software maintainability and the metrics. They have detected the maintainability in software using OO metrics. Majumder et al. [13] have taken a software jfreechart, to perform a case study in order to propound a model for maintainability prediction. Padhy et al. [14] have proposed a novel technique to evaluate software metrics using an evolutionary intelligence method. The proposed model computes threshold values on the software metrics parameters using the proposed method. The literature review provides an existing model for detecting the maintaining the reliability in OO software.

Objective of the proposed method: to analysis, the sub-module implementation as software ontology for opens sources software. The open software has many bugs in the quality [17, 18]. This research chapter explains how to reduce bugs that are reported as issues of the software. The motivation for the proposed concept is based on the literature review. A software bug called fault means an event occurs when delivered services are diverted from the existing services. This study mainly focuses on:

- How software maintainability change?
- What are the fundamental issues of declining quality of software?

The maintainability issues and their hierarchy are described in Fig. 1. Many research questions are analyzed in this process, like classification of bugs and bugs process and their issues [20]. This study influences the bugs' response between client and server responses. During the maintainability process, the sub-modules of software which focus on software quality are dependent on each component of modules and also identify root causes of bugs along the chain process (see in Fig. 1).

The software hierarchy and upper level focus on business classes like costing and timing process of software that's value proposition depend on effective services,

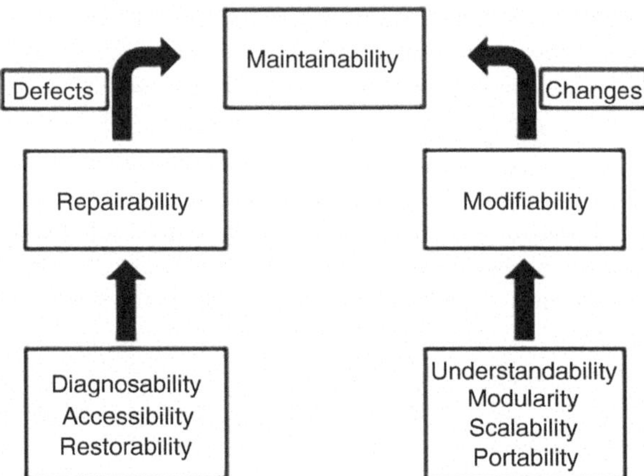

Fig. 1 Software quality maintainability process

maximum utilization of resources, module dependency, and flexibility of the software [21]. The next level of software maintainability monitors the different tasks of software, like critical and complex stage of software quality, which is the main cause of various maintainability relationships.

The propose model handles the defects of software such as diagnosability which means to identify the error messages (log messages, failure of test, unreliable information for proper assessment) which are the main cause of software quality. Another important factor is accessibility which means that the user is able to access the needed domain of software [22–25]. The cause of accessibility is log messages and test failure on the user accessing data and redirection of function processing or unaccepted location of data place.

3 Pre-requisite Knowledge

This section describes some relevant knowledge regarding the experiments performed on the proposed model and also measures the effectiveness of the proposed method.

3.1 Random Forest

The software provides auto-guided learning technique, which finds the class of the test data by aggregating the decisions of the decision tree classifier applied on the random set of training data (see Fig. 2).

The reason behind the selection of this classifier is its capability of handling large datasets by predicting more accurate results. Moreover, its effectiveness for missing data estimation puts it into the nomination of classifiers.

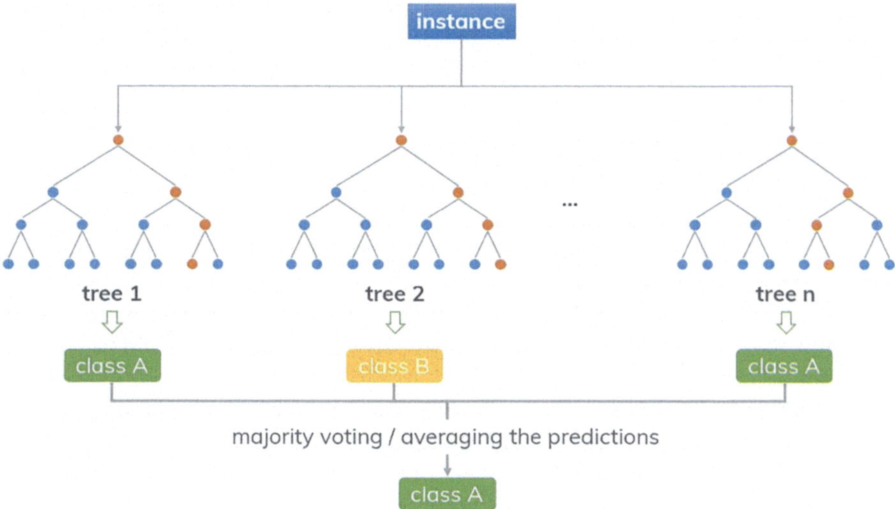

Fig. 2 Random forest

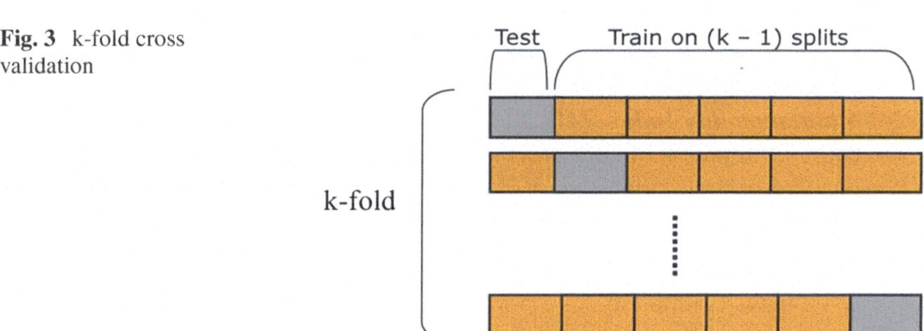

Fig. 3 k-fold cross validation

3.2 K-Fold Cross Validation

In *k*-fold cross validation, the sample data is split into k subsets, which are called folds. So, ML model implements on all split data subset and processes, each one (*k*-1) of the subsets in the sequence as a trailed till kth times (see Fig. 3).

The proposed model evaluates each training performance using fivefold cross validation.

3.3 Software Metrics

Chidamber and Kemerer (CK) [15–17] have proposed some OO design metrics to compute the maintainability effort. Some of them are discussed in this section [24].

- **Size 2**: It measures the size of the class and provides valid information of sub-modules and details of attributes in a class.
- **Size 1**: The preliminary parameters of the class are listed below.
  - ***Number of Local Methods (NOM):*** *This process describes static or local methods/functions in object-oriented matrix.*
  - ***Data Abstraction Coupling (DAC)***: The data format and its retrieval without changing internal working process during data access by the application.
  - ***Message Passing Coupling (MPC)***: This parameter provides log details of message variables and send statement which are define inside the class.
  - ***Lack of Cohesion on Methods (LCOM)***: The cohesion parameter is used to link the sub-models or local method properly, and data flow should be smooth during testing.
  - ***Weighted Methods per Class (WMC)***: These parameters compute the possible bugs based on existing variables using Mccabe's cyclomatic principal.
  - ***Response for a Class (RFC)***: These parameters provide the relationship and flow of data during debugging and testing to measure the cardinality of the respective class responses.
  - ***Number of Children of a Class (NOC)***: It provides information related to all sub-models in each class.
  - ***Depth of Inheritance Tree (DIT)***: The file and directory creation and analysis process related to each class.
  - ***Maintainability Index (MI)***: Software metric which computes and manages entire information of each process like source code. The MI is computed as a factored formula consisting of Halstead volume, Cyclomatic complexity, and lines of code.

HV = Halstead volume
CC = Cyclomatic complexity
LOCO = number of source lines of code

From these measurements, the MI can be calculated by Eq. (1)

$$M = \max(\mathrm{o},(175 - 5.3 * \left(i_n\left(\mathrm{HV}\right) - 0.23 - 16.2 * i_n\left(\mathrm{LOCO}\right)\right) * \frac{100}{171} \tag{1}$$

3.4 Performance Assessment Measures

Accuracy: It can be computed by dividing the true predicted results by the total results. In terms of confusion matrix data, accuracy can be calculated using the following formula:

$$\mathrm{ACC} = \frac{\mathrm{TN} + \mathrm{TP}}{\mathrm{FP} + \mathrm{TP} + \mathrm{FN} + \mathrm{TN}} \tag{2}$$

F-score: F-score is computed and analyzes the performance of the proposed model which provides information for precision and recalls scores as a single measure. It can be calculated using the following formula:

$$\text{Fscore} = \frac{2 * P_{rc} * \text{Rec}}{P_{rc} + \text{Rec}} \tag{3}$$

3.5 Need of Predictability

The software quality is fully dependent on the software development process attributes like accuracy and proper meaningful documentation, reliable modulation, and proper cohesion implementation [25]. The software maintainability focuses on the maximum costing of project budget. The evaluation of software is also part of same model like maintenance after deliver to the client, during the operation mode, many changes are required by the client, which are considered as unpredictable changes due to a variety of faults, scope which increase the cost of software maintenance after delivery to the client.

The software volumes are automatically started here with risk analysis process [19]. These stages indicate that the current design of the software is not much reliable under the current conditions, so software moves to the prediction process and development phase, which is also called stress testing. The same recycle is continuing until clients are satisfied.

4 Proposed Methodology

4.1 Datasets

The reliability of sample dataset is a very important factor for consolidated performance analysis. Our results are based on authentic dataset [18–22]. Research institutes are providing open access for researchers the profiled dataset based on various parameters (see in Table 1).

Table 1 Dataset sample

Group	Dataset	Total number of instances	No. of defective instances	No. of non-defective instances
SOFTLAB	ar3	63	8	55
SOFTLAB	ar4	107	20	87
SOFTLAB	ar6	101	15	86

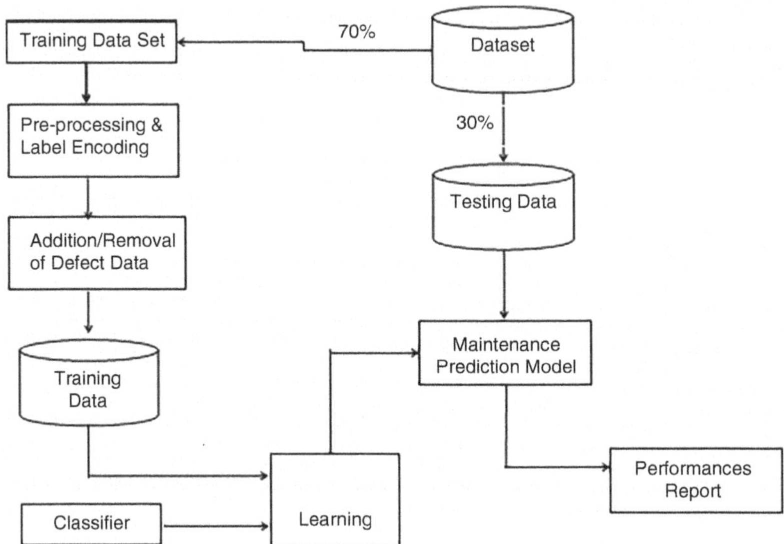

Fig. 4 Proposed methodology

4.2 Proposed Method

In this methodology, the research dataset is divided for the client testing process or white box testing on various parameters through various evaluation processes on the data ratio (7:3). Data preprocessing is applied along with label encoding. We have computed the maintainability index (MI) using the stated formula. In order to determine the impact of defects in the maintainability prediction, we have classified it into a binary class problem. MI greater than 20 depicts high maintainability and is coded as class 1, and less than or equal to 20 has low maintainability and is coded as class 0 (see Fig. 4).

We conducted two experiments: In the first experiment, we have predicted the MI using the dataset without the number of defect attribute. In the second experiment, we have predicted the MI target class with the number of defect attributes along with the other metrics. The proposed model used the random forest principle for predicting the ensemble learning classifier. Software performance is evaluated using accuracy and *F-score* as the performance measures. The below figure describes the proposed methodology.

5 Results

This section shows the results that are obtained through our experiments using the proposed method.

Fivefold Cross Validation

The proposed model is applied fivefold cross validation process to analyze the client training performance on the software. In this case, the proposed method did not over fit during training, the various test on the number of folds and discuss in five fold cross validation outcomes and performance of model training under five fold cross validation. The proposed model provides reliable services on the given set of training data gaining the highest training accuracy, which is 85.92%.

Table 3 shows the performance assessment results of the model that is trained using the dataset that does not include the number of defect attribute during the training of the classifier.

Table 4 shows the performance assessment results of the model that is trained using the dataset that includes the number of defect attributes during the training of the classifier.

The results obtained from the two tables show that the software defect has a great impact on the software maintainability predictability. The results obtained using case 2 more accurately predict the software maintainability index as compared to the case 1 when the defects are not taken into consideration during the training of the model. Moreover, we have compared the two cases using the graphical method. The results obtained in terms of accuracy are shown in Fig. 5. It shows that the

Table 2 Results of fivefold cross validation

No. of folds	Accuracy (%)
1	85.12
2	84.92
3	85.17
4	85.42
5	85.16

Table 3 Results of MI prediction without the number of defects attribute

	Accuracy		
Dataset	Training	Testing	F-score
AR3	0.86	0.81	0.83
AR4	0.89	0.85	0.78
AR6	0.78	0.80	0.75

Table 4 Results of MI prediction with the number of defects attribute

	Accuracy		
Dataset	Training	Testing	F-score
AR3	0.88	0.86	0.88
AR4	0.89	0.87	0.84
AR6	0.77	0.86	0.82

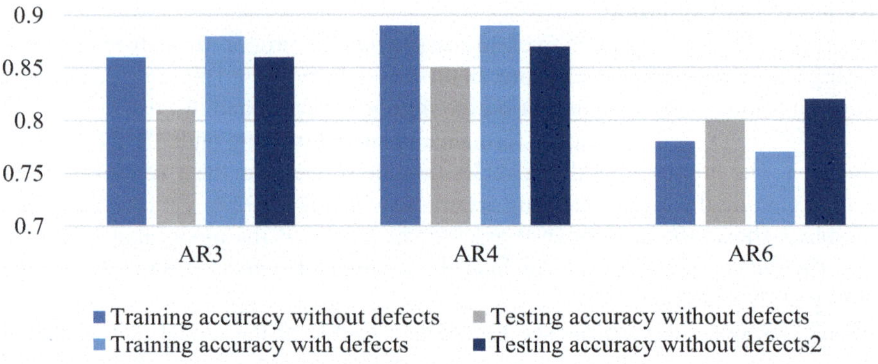

Fig. 5 Comparison of two proposed maintainability prediction techniques using accuracy

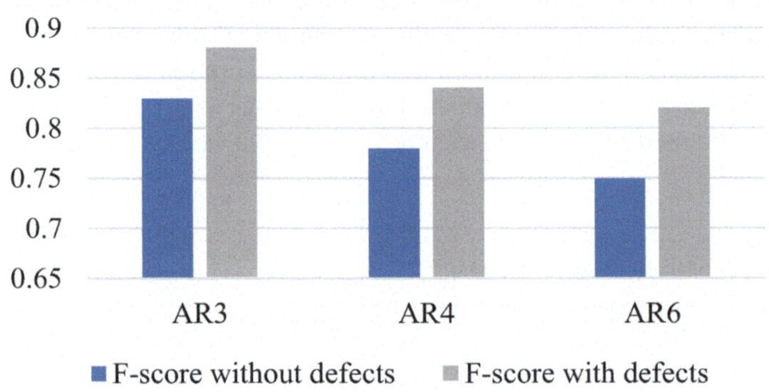

Fig. 6 Comparison of two proposed maintainability prediction techniques using F-score

training and testing accuracy thus obtained is higher when defects are taken into consideration as compared to the results in case 1.

Similarly, the F-score measure obtained is compared in Fig. 6. It shows that the model when feed using the defect information is well trained and hence well predicts the software maintainability as compared to that case when the defect data is not available.

6 Conclusion

The major objective is to compute the effect of defect prediction data on software maintainability detection. The experimental results performed using the three datasets AR3, AR4, and AR6 have proven that the MI prediction method gives better performance based on F-score when the defects are taken into consideration to train the random forest classifier, than the trained classifier without using the defect information.

Thus it can be concluded that the defect prediction is useful if the maintainability of the system is to be detected well resulting in production of an error free and reliable software as the defects information aids the classifier in better learning by overcoming the over-fitting issue. Therefore better the learning better the prediction. In the future, other huge datasets can be also considered with more defects and bugs information to detect the software maintainability.

Acknowledgments This research is funded by Vietnam National Foundation for Science and Technology Development (NAFOSTED) under grant number 102.03-2019.10.

References

1. Goel, S., Krishnamurthy, S., & Hampsey, M. (2012). Mechanism of start site selection by RNA polymerase II: Interplay between TFIIB and Ssl2/XPB helicase subunit of TFIIH. *Journal of Biological Chemistry, 287*(1), 557–567.
2. Ghosh, S., Dubey, S. K., & Rana, A. (2012). Fuzzy maintainability model for object oriented software system. *International Journal of Computer Science Issues (IJCSI), 9*(4), 338–342.
3. Zhang, W., Huang, L., Ng, V., & Ge, J. (2015). SMPLearner: Learning to predict software maintainability. *Automated Software Engineering, 22*(1), 111–141.
4. Ye, F., Zhu, X., & Wang, Y. (2013). A new software maintainability evaluation model based on multiple classifiers combination. In *2013 International Conference on Quality, Reliability, Risk, Maintenance, and Safety Engineering (QR2MSE)* (pp. 1588–1591).
5. Lin, M. J., Yang, C. Z., Lee, C. Y., & Chen, C. C. (2016). Enhancements for duplication detection in bug reports with manifold correlation features. *Journal of Systems and Software, 121*, 223–233.
6. Jha, S., Kumar, R., Abdel-Basset, M., Priyadarshini, I., Sharma, R., et al. (2019). Deep learning approach for software maintainability metrics prediction. *IEEE Access, 7*, 61840–61855.
7. Oman, P., & Hagemeister, J. (1994). Construction and testing of polynomials predicting software maintainability. *Journal of Systems and Software, 24*(3), 251–266.
8. Welker, K. D. (2001). The software maintainability index revisited. *CrossTalk, 14*, 18–21.
9. Aggarwal, K. K., Singh, Y., Kaur, A., & Malhotra, R. (2008). Application of artificial neural network for predicting maintainability using object-oriented metrics. *World Academy of Science, Engineering and Technology, International Journal of Computer, Electrical, Automation, Control and Information Engineering, 2*(10), 3552–3556.
10. Thwin, M. M. T., & Quah, T. S. (2005). Application of neural networks for software quality prediction using object-oriented metrics. *Journal of Systems and Software, 76*(2), 147–156.
11. Zhou, Y., & Leung, H. (2007). Predicting object-oriented software maintainability using multivariate adaptive regression splines. *Journal of Systems and Software, 80*(8), 1349–1361.
12. Zhou, Y., & Xu, B. (2008). Predicting the maintainability of open source software using design metrics. *Wuhan University Journal of Natural Sciences, 13*(1), 14–20.
13. Majumder, R., Som, S., & Gupta, R. (2017). Vulnerability prediction through self-learning model. In *2017 International Conference on Infocom Technologies and Unmanned Systems (Trends and Future Directions) (ICTUS)* (pp. 400–402).
14. Padhy, N., Panigrahi, R., & Neeraja, K. (2021). Threshold estimation from software metrics by using evolutionary techniques and its proposed algorithms, models. *Evolutionary Intelligence, 14*, 1–15.
15. Chidamber, S. R., & Kemerer, C. F. (1991). Towards a metrics suite for object oriented design. In *Conference proceedings on object-oriented programming systems, languages, and applications* (pp. 197–211).

16. Chidamber, S. R., & Kemerer, C. F. (1994). A metrics suite for object oriented design. *IEEE Transactions on Software Engineering, 20*(6), 476–493.
17. Chidamber, S. R., Darcy, D., & Kemerer, C. F. (1998). Managerial use of metrics for object-oriented software: An exploratory analysis. *IEEE Transactions on Software Engineering, 24*(8), 629–639.
18. Ghenname, M., Abik, M., Subercaze, J., Gravier, C., Laforest, F., & Ajhoun, R. (2015). Hashtag-based learning profile enrichment for personalized recommendation in e-learning environments. Int Rev Comput Softw (IRECOS), 10, 891–899.
19. Gupta, S., & Sharma, K. P. (2020, March). A review on applying tier in multi cloud database (MCDB) for security and service availability. In *2020 International Conference on Computer Science, Engineering and Applications (ICCSEA)* (pp. 1–4). IEEE.
20. Solanki, M. S., Goswami, L., Sharma, K. P., & Sikka, R. (2019, December). Automatic detection of temples in consumer images using histogram of gradient. In *2019 International Conference on Computational Intelligence and Knowledge Economy (ICCIKE)* (pp. 104–108). IEEE.
21. Le, B. N., Le, D. N., & Nguyen, G. N. (2016, November). Optimizing selection of PZMI features based on MMAS algorithm for face recognition of the online video contextual advertisement user-oriented system. In *International symposium on integrated uncertainty in knowledge modelling and decision making* (pp. 317–330). Springer.
22. Le, D. N., Nguyen, G. N., Bao, T. N., Tuan, N. N., Thang, H. Q., & Satapathy, S. C. (2021, April). MMAS algorithm and Nash equilibrium to solve multi-round procurement problem. In *Advances in systems, control and automations: Select proceedings of ETAEERE 2020* (pp. 273–284). Springer Singapore.
23. Bao, T. N., Huynh, Q. T., Nguyen, X. T., Nguyen, G. N., & Le, D. N. (2020). A novel particle swarm optimization approach to support decision-making in the multi-round of an auction by game theory. *International Journal of Computational Intelligence Systems, 13*(1), 1447–1463.
24. Le, D. N. (2017). A new ant algorithm for optimal service selection with end-to-end QoS constraints. *Journal of Internet Technology, 18*(5), 1017–1030.
25. Le, D. N., Nguyen, G. N., Garg, H., Huynh, Q. T., Bao, T. N., & Tuan, N. N. (2021). Optimizing bidders selection of multi-round procurement problem in software project management using parallel max-min ant system algorithm. *Computers, Materials & Continua, 66*(1), 993–1010.

EncryptoX: A Hybrid Metaheuristic Encryption Approach Employing Software Testing for Secure Data Transmission

Sushree Bibhuprada B. Priyadarshini, Aayush Avigyan Sahu, Vishal Ray, Padmalaya Ray, and Swareen Subudhi

1 Metaheuristics and Background Study

Nowadays, metaheuristic is a very popular approach that involves analytical optimization while incorporating higher level procedures. In the current chapter, we have used a hybrid metaheuristic strategy, namely, EncryptoX, that employs software testing for secure data transmittal [1].

1.1 Introducing EncryptoX

Cloud technology and cybersecurity are the emerging topics of today's era of technological progression [3]. In the current chapter, we are pleased to present the novel project EncryptoX, where proper planning and metaheuristic methods employing software testing were followed while doing this project so as to make it efficient. Practical knowledge is as important as theoretical knowledge while testing our proffered system.

1.2 Objective

The main aim is to achieve a secure platform for storing of files on cloud using hybrid cryptography. The predominant issue of data security and privacy is increasingly affecting the readiness of small and medium businesses to migrate their data from on-site to cloud facilities. The cloud technology has improved drastically in

S. B. B. Priyadarshini · A. A. Sahu (✉) · V. Ray · P. Ray · S. Subudhi
Siksha 'O' Anusandhan Deemed to be University, Bhubaneswar, Odisha, India

M. Khari et al. (eds.), *Optimization of Automated Software Testing Using Meta-Heuristic Techniques*, EAI/Springer Innovations in Communication and Computing, https://doi.org/10.1007/978-3-031-07297-0_11

the past several years, and users can easily access data from cloud remotely with unlimited access.

There is a very big concern whether such providers are able to provide a secure and reliable service which can keep the user's files safe and secure from hackers. Appreciably, some have managed to deploy either symmetric or asymmetric cryptography techniques to achieve some level of security on cloud storage. This project focuses on cloud storage security issues while giving particular attention to emerging hybrid cryptography [1, 4].

Cloud computing is an innovative model for delivering services and information using current technologies. Physical storage devices are gradually becoming obsolete. The globalization of business that has necessitated sharing data for working collaboratively and on multiple personal devices is the reason for this change. Cloud storage is most applicable in the new era because it facilitates easier collaboration and convenient shifts from one device to another by providing a singular platform to connect multiple individuals and devices remotely [4, 5]. However, cloud storage technologies yet introduce various data storage securities such as leakage, unwarranted access, and illegal modification. Such risks have necessitated the implementation of hybrid cryptography and other techniques of ensuring data on cloud facilities to be secure.

The implementation of hybrid cryptographic techniques is better than implementing either symmetric or asymmetric cryptography while employing metaheuristic scheme. In the analysis of cloud storage security, the hybrid cryptography is very secure and reliable to secure the data and in keeping it safe from hackers [2, 3]. The techniques are achieved through mechanisms of access control, authorization, authentication, and confidentiality. Overall, the objective of our project is to provide a secure platform where user can store their data and information without having any second thoughts and can secure their data encryption methods [4, 5].

2 EncryptoX and Cloud

Currently there is a large movement in the cloud computing technology to mainstream processes and improve the current methods of storing data online. As in the case with any file sharing, this effort is difficult without an "information at your fingertips" type of system application. A well-designed database for this is a large improvement over the current methods of managing files and allowing quick responses to user requests for accessing their own data [1, 2].

Several problems with the current case file storing management system have been identified. The current system would benefit from a centralized user record repository and communication platform in which all user records are kept and from which the state of access and security is low. In the cloud computing, resources are shared among all the servers, users, and individuals. So it becomes difficult to keep a check on the security of the files of users.

As a result, it somehow becomes easier for a hacker to access, misuse, and destroy the original form of data. In case of compromise at any cost, the data of users are compromised. Therefore, there is a need for a robust secure technique for data security that becomes vital. So we need to come with a more secure technique which can be done using hybrid cryptography to secure our file storing management system.

2.1 Specifications

Cloud computing is one of the most popular technologies in the world with it being used in many fields such as defense, healthcare, industries, institutions, etc. in storage of bulk data. The biggest advantage of cloud is that users can access their files remotely without directly accessing the servers. However, the security of data stored online have raised concerns over the years; therefore, new techniques are being developed to keep our files and other vital information secure.

One of the most used technique is the use of hybrid cryptography algorithm. By implementing hybrid cryptography and metaheuristic so that the system becomes very secure, even if there is a security breach, the attacker will only access the encrypted data which is not human readable. Hybrid cryptography utilizes the AES and RSA algorithms, where the key is encrypted by RSA and the data is encrypted by AES. The main motto behind this project is to provide an application which uses hybrid cryptography to keep the user files secure and stop them falling prey into the hands of hackers and eliminating the man-in-the-middle attack [2, 3].

Once the EncryptoX website is loaded, the user can choose between the "upload" and "restore" button. If the user clicks on the upload button, then it gets navigated to the "Upload" page where the user can select a particular file and upload it in the storage platform, and then it gets navigated to the success page where the user can download the public key to access the file to restore. If the user clicks on the "restore" button, then it gets navigated to the restore page where the user then gets an option to choose between going back to home page or the download key page. Finally, the user gets navigated to download page and hence can download the file. Thus successful completion of file storing is done sequentially.

2.2 Hardware Specification

As it is a computer application, there is a minimum hardware specification required to run the application smoothly, such as the following:

(i) Minimum 2 GB of RAM.
(ii) 256 of storage space for the application
(iii) Intel core i3/AMD Ryzen 3 or above.

2.3 *Software Specification*

Following are the software specifications employed for our proffered software testing Scheme.

(i) A web browser which supports HTML5 and JavaScript.
(ii) A Python IDE like PyCharm.
(iii) XAMPP server to host the web application.
(iv) Flask virtual environment to run the application.

3 Analyzing Existing System

Day by day, there is an increase in cyberattacks, and data breach is happening everywhere, and it's high time that we take a step forward in cybersecurity. This section mainly deals with the comparison between the existing applications and our application. After some thorough research and study, we found that there are many research papers on this topic and lots of research is going on. But there is no software application on this available for public use. Few prototypes are available on the internet, and there are some ways in which our project is different from the other, and in this section, we have thoroughly gone through the comparison among other application [4, 5].

3.1 *Existing System*

A user is an entity who has some data to be stored in any file storage system. Suppose the user wants to store the files on any cloud storage system, first the user will choose the files to be uploaded, and then the user will upload them on the cloud, but there are high chances of man-in-the-middle attack where the files of user are compromised by the hacker even before it is uploaded on the cloud. As there was no encryption of any type, the hacker can easily access the user files and can use it for himself, thus making the traditional system flawed. There are certain limitations we face in the existing system such as the following:

- Cloud service providers should make the security of user data as their main priority.
- Service availability failure and the possibility of malicious software and hackers in the cloud infrastructure.

3.2 Proposed Testing System

Currently there is no such system available. There are many theoretical models and prototype with very limited user interface. Our project provides new model greatly in terms of user interface and with the encryption that helped in the removal of the man in the middle attack [1–5] as follows:

(a) We have implemented the encryption and decryption techniques and algorithms to provide security to our cloud platform.
(b) Secure upload and download of file transfer is proposed here.
(c) Easy and user-friendly website.
(d) Responsive website with secure data storage.

3.3 Feasibility Study

The project "EncryptoX-Secure Storage System using Hybrid Cryptography" aims at encrypting the data inside the file such that no new user can see it. Firstly, we made this approach technically possible by implementing AES and RSA encryption algorithm technique and cryptography. Secondly, our project will also make a huge impact on the society as it will be very profitable to technocrats, professionals, and common people as it is free of cost.

Secondly, our proffered strategy is also legal as it abides with all cyberlaws and norms. Thirdly, our proposed approach will also support the changes as it will also successfully encrypt images and videos, and lastly, the project will also be delivered in time as the file will be uploaded in very less time and also the encrypted file will be downloaded in a very less time. Recent studies have also ensured that hybrid encryption technique has been increasing day by day.

4 Software System Analysis and Design

With the rise in cybersecurity attacks and data breach of many famous companies' database, it becomes high time that we take some strict actions against it. So in order to achieve data security, we made EncryptoX [6], a simple web application that encrypts the files of users and makes it secure against cyberattacks. The user gets a key after encrypting the files, and only that valid logical key can decrypt and restore the files into its original form. Therefore while designing this project we have kept two things in mind:

A. Usability.
B. User interface.

To achieve the abovetwo we had to design the application in such a way that not only it would be simple to use but also it becomes very intuitive. The back end was to be made first, for that we used Python and some python libraries and packages, namely, Flask, Werkzeug, cryptography, and ChaCha20 encryption algorithm. ChaCha20 encryption and decryption algorithms and codes were available online so that we directly implemented them in our application code [6].

The back-end codes are divided into many parts, namely, application, the main application code which calls the functions and other code at the time when snippets called; encrypter, the encryption code which encrypts the files and returns an encrypted file and a key; decrypter, the decryption code which decrypts the files with the help of the key; and restore, which helps to restore the files into its original form and tools and which has packages and codes required for the application [7]. The user interface is designed using HTML, CSS, and JavaScript. The user interface is kept very user friendly and intuitive for users so that they don't face any problem while using the application [4, 5, 8].

4.1 Requirement Specification

Two types of specifications are employed here as follows:

4.1.1 Functional Requirement

The functionality of this project is to encrypt user files so that it can be secured from cyberattacks and data breaches. When the application is run, it invokes the index page which has two options—upload or restore. Upload enables the user to encrypt the files and upload them into any storage system either into system or into any cloud infrastructure. The files can only be decrypted using the valid logical key, and if by chance the encrypted files fall prey to any cyberattack, then the hacker will only get the encrypted file which will not be in human-readable format [7, 8].

(a) **Upload:** The user will have the option to upload the files into the storage application by clicking on this button. When the user clicks on these buttons, he will be prompted to choose a file which then will be encrypted and stored in the uploads folder of the storage system, and a key will be generated.
(b) **Download Key:** The user will download the key which can only decrypt and restore the files.
(c) **Home:** This button enables the user to get back to home page.
(d) **Restore:** When the user clicks on this button, he is to be asked to choose the file that he wants to decrypt and restore. With the help of the key, the file is decrypted and restored, and then the user can download the restored file.

4.1.2 Nonfunctional Requirements

(a) **Scalability:** The application is able to perform the desired actions of encryption and decryption within a few seconds with a hassle-free user experience.
(b) **Reliability:** EncryptoX uses hybrid encryption techniques, namely, AES for data and RSA for key, which is one of the most reliable encryption techniques used for the assurance of privacy of data.
(c) **Robustness:** The application is quite robust due to its straightforward user interface and its implementation.
(d) **Consistency:** The application is consistent throughout the process. The application works well or performs well as expected every time.
(e) **Usability:** This is a web application, and the users must know how to use web browser and how to run a web application. As the user interface is very user friendly, the users won't find any difficulty while using the application.

4.2 Design Steps and Criteria

The main focus during making of this project was to give the user a hassle-free experience while using the application. For this, the front end was designed accordingly so that user will find the application very intuitive to use [8, 9].

The first step toward making this project was to design and code the back end. The back end is completely coded in Python with many Python libraries used like Flask, cryptography, and WSGI. OOPS concept is used with the application coding so that the codes run without any problem. The front end is written in HTML, CSS, and JavaScript, and some theme packages are used to make the front end look better and user-friendly [10]. There are some design constraints:

(i) It requires a latest web browser which supports HMTL5 and JavaScript.
(ii) The system on which this application is running should have python installed along with all the necessary libraries and functions.
(iii) XAMPP server is required to run this application as it's a web application which is supposed to run a web server.
(iv) It can also be integrated and run on a cloud server service like AWS to provide more practical experience, but for that then we have to pay monthly payments to AWS for hosting and providing other services.

5 Algorithms and Pseudo Code

The main algorithm used in our application is ChaCha20—Poly1305 encryption. This algorithm is used to encrypt and decrypt the files of the user. This is one of the most reliable encryption algorithms available to us. Its pseudo code is as follows:

The basic operation of the ChaCha algorithm is the quarter round. It operates on four 32-bit unsigned integers, denoted a, b, c, and d.

The operation is as follows (in C-like notation):

1. a += b; d ^= a; d <<<= 16;
2. c += d; b ^= c; b <<<= 12;
3. a += b; d ^= a; d <<<= 8;
4. c += d; b ^= c; b <<<= 7;

where "+" denotes integer addition modulo 2^32, "^" denotes a bitwise Exclusive OR (XOR), and "<<< n" denotes an n-bit left rotation (toward the high bits).

For example, let's see the add, XOR, and roll operations from the fourth line with sample numbers:

a = 0x11111111
b = 0x01020304
c = 0x77777777
d = 0x01234567
c = c + d = 0x77777777 + 0x01234567 = 0x789abcde
b = b ^ c = 0x01020304 ^ 0x789abcde = 0x7998bfda
b = b <<< 7 = 0x7998bfda <<< 7 = 0xcc5fed3c

For a test vector, we will use a ChaCha state that was generated randomly:

Sample ChaCha state

879531e0 c5ecf37d 516461b1 c9a62f8a
44c20ef3 3390af7f d9fc690b 2a5f714c
53,372,767 b00a5631 974c541a 359e9963
5c971061 3d631689 2098d9d6 91dbd320.

We will apply the QUARTERROUND(2,7,8,13) operation to this state. For obvious reasons, this one is part of what is called a "diagonal round":

After applying QUARTERROUND(2,7,8,13).

879531e0 c5ecf37d *bdb886dc c9a62f8a
44c20ef3 3390af7f d9fc690b *cfacafd2
*e46bea80 b00a5631 974c541a 359e9963
5c971061 *ccc07c79 2098d9d6 91dbd320.

Note that only the numbers in positions 2, 7, 8, and 13 changed.

6 Testing Process

6.1 Unit Testing

It is the testing of all the units of the application individually to check whether each unit is free of bugs and giving the desired results or output [11, 12]. Our application back-end codes are divided into five units or parts, namely, app.py, encrypter.py, decrypter.py, restore.py, and tools.py. The unit testing was carried out by testing each unit codes to check whether they are free of bugs and not throwing any exception [13–16]. While performing the unit testing, all the units came to be error free with few warnings. The main application program was error free with no warnings. The other units were also found to be error free with two unit codes throwing three to four warnings, which was expected due to implementation of different algorithms and Python packages in the application codes.

6.2 Integration Testing

This is the testing phase where all the units are combined or connected, and the main application codes are thoroughly tested. During the integration testing we had to implement the front end to check whether the codes were working properly or not [14, 15, 17]. The front end during this testing was developed in basic HMTL with simple buttons to check the working of the application. Figures 1 and 2 are the screenshots of the output. Figure 3 shows the screenshot of success stage of unit testing output.

Fig. 1 Index page unit testing

Fig. 2 Upload page unit testing

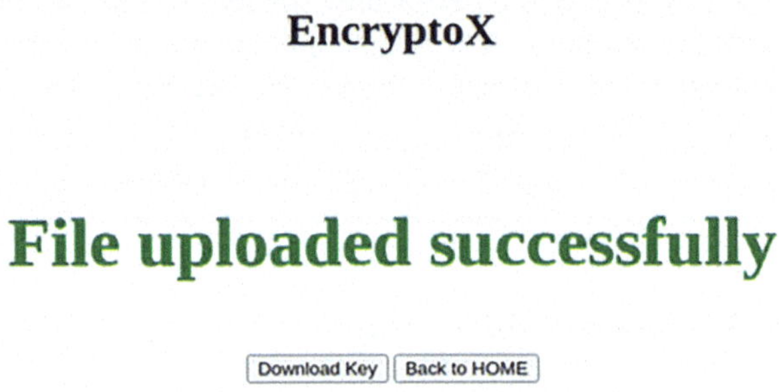

Fig. 3 Success page unit testing output

7 Results and Discussion

This section deals with the output of the application. Also the user interface was our main focus while developing this application. The output of our application depends on the user's choice. If the user chooses to upload the file, then the output which we get is the encrypted file along with the key to decrypt and restore it into its original form. If the user chooses to restore the file, then he gets the file decrypted and restored into its original form [3, 16].

7.1 User Interface

The main focus while making this application was given to the user interface. The user interface of the application is very intuitive and simple to use so that the user can have a hassle-free experience while using the application. Figures 4, 5, 6 and 7 illustrate the user interface.

7.2 Results and Discussion

The output of this application depends on its action chosen by the user. If the user chooses to upload a file, then it will return the encrypted file along with the key required to decrypt and restore it into its original form. If the user chooses

Fig. 4 Index page

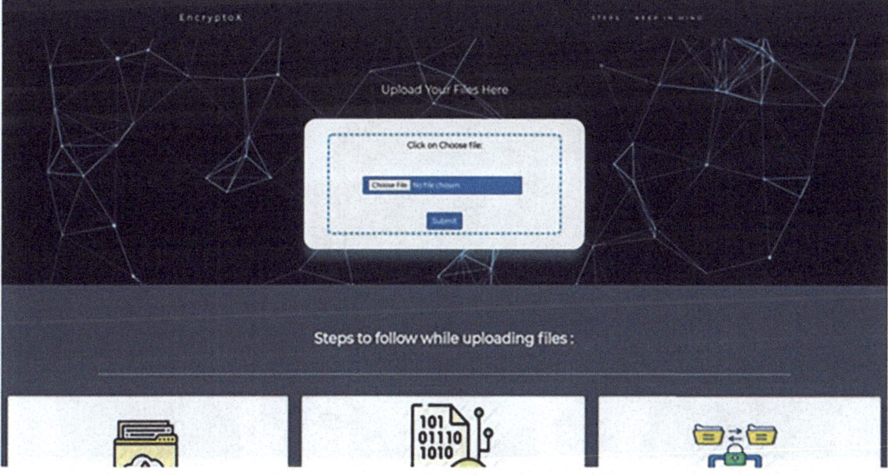

Fig. 5 Upload page

to restore the option, then the output will be the decrypted file restored into its original form.

8 EncryptoX Facilities

As the technology is changing everyday, our application is also required to evolve in order to cope up.

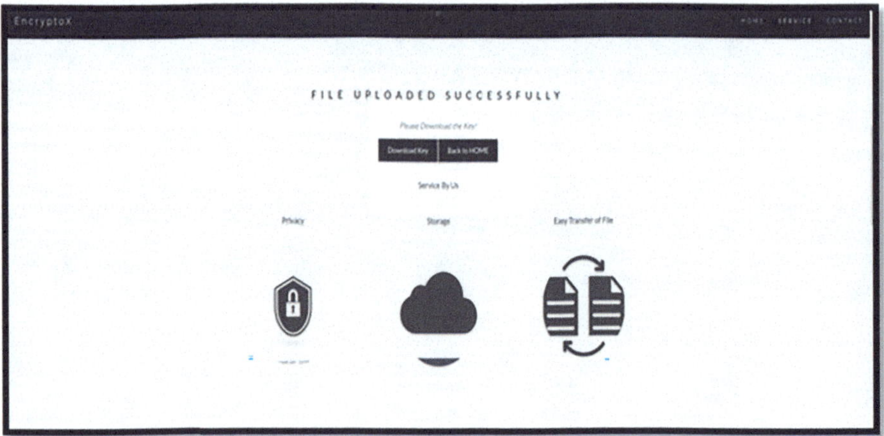

Fig. 6 Success page

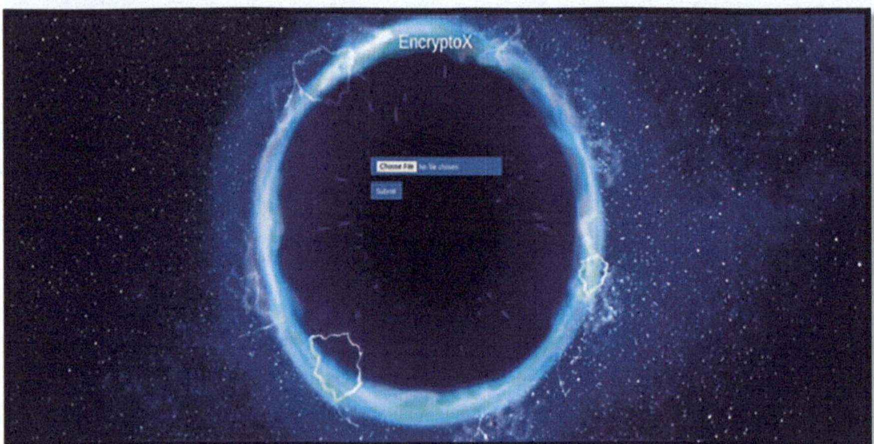

Fig. 7 Download page

(a) **The user sign in–sign out option:** There will be sign in and sign out option for user, so that when not required the user can simply sign out of the application. This feature will make the application a bit more secure and conventional for mass use.
(b) **A mobile application version:** As majority of people use their smartphones to do their majority of day-to-day work, so it will be more conventional to make an Android/IOS EncryptoX application so that people can use it more efficiently.
(c) **A GUI-based application software:** This application is a web application and requires specific python libraries and virtual environment to run but a fully develop GUI software which has all the libraries and required function encap-

sulated just like other regular conventional software which is only required to install once, and it will take care of rest.

(d) **Option to switch between storage system:** The user will get the option to choose a storage system which is registered, such as local storage system, cloud storage, or even flash drives.

9 Socioeconomic Impact

Basically, the EncryptoX represents the app for both MAC and windows that permits the encryption file prior to sending it. Its main objective is to protect and secure the data.

9.1 Practical Relevance

With the rise in cybercrimes and data breaches, now it's high time we take serious and necessary steps to keep our data secure. Users keep their data in different storage systems like their local storage on their devices or on the cloud. The main problem with all these storage systems is that it stores data in its original form, so if by chance any hacker hacks the system and gets his hands on the files, they can easily misuse.

To avoid such attacks we have tried to make and application named EncryptoX, which can encrypt user data before storing it on any file system and can decrypt and restore it into its original form with the help of the valid key. This project is very practical and can be of great use for people to keep their files secure. This is a simple web application which can run on any computer system with python installed and by running it on a virtual environment. It's a very easy-to-use application with an intuitive user interface. It can easily be used for local file storage system.

9.2 Global Impact

In the twenty-first century, digitalization is increasing rapidly, and with this increase there is also a rapid increase in data leaks and data breach. Recently we have seen that a huge amount of data has been leaked from the Domino's and Air India's database. Our data can be only safe through encryption. So our project "EncryptoX being a Secure Storage System using Hybrid Cryptography" will be a game changer in the field of encryption and cryptography. The current chapter will make a huge impact globally by safeguarding the sensitive data of user from unknown users. Most often it is seen that people do not remember bank account number, email addresses, passwords, etc. So with our project people can safely store there bank account number and password in file such that no one can know them. Basically, it will encrypt all passwords, email addresses, and bank account numbers which are present in the file.

9.3 Lifelong Learning

In the modern era, data is the new oil, and many hackers try to steal sensitive data from the users, which is a great threat. So, we all should take a step forward to protect our data. In this chapter, we have secured our data through encryption [2, 5]. We can also secure our data from hackers through many ways such as the following:

Firstly, by using firewall that will avert unauthorized access to the network and notify the user of intrusion attempts.

Secondly, by installing antivirus software as it provides many features and security to protect the system by detecting real-time threats to ensure the data becomes safe.

Thirdly, by setting complex passwords—the password must be at least eight characters and should have combinations of numbers, uppercase and lowercase alphabets, and computer symbols.

Fourthly, by the using two-factor authentication, which hackers will find it difficult to hack the data as two-factor authentication will act as second layer protection.

Lastly, by clearing browsing history, cookies, and cached files.

10 Conclusion and Future Work

In this digital modern world where the user data plays a very vital role, we must keep our files and data secure. Nowadays, with increase in cybercrimes and data leak, it's high time that we use applications which are very secure and can keep our data secure. People have many files and data to store, and sometimes device storage falls short for the user, so the user many times opts to store their data on cloud systems to save space on their devices. With the growing technology and faster development in the society, we need to advance and enhance the encryption methods. To keep data secure from hackers, many countries and companies are working on new techniques and methods. People's information is a big responsibility; thus the world is moving forward to digital era.

Hybrid cryptography is a very new technique with many benefits. It's better than using a single encryption technique and keeps the data more secure. In today's digital world where data is the most precious thing about an individual on the internet, it's very important to keep the user data and files secure. In order to achieve this, we tried to make a metaheuristic application, namely, EncryptoX, which attempts to fulfill all these criteria for data security. It's a simple application and easy to use with a very intuitive user interface. It's now time that we as an individual understand the importance of data and keep it secure from our side rather than relying on the service providers.

References

1. Osman, I., & Kelly, J.. (1996). *Meta-hueuristics theory and applications*. ISBN: 978-1-4613-1361-8, pp. 1–690.
2. Sharma, M., & Kaur, P. (2021). A comprehensive analysis of nature-inspired meta-heuristic techniques for feature selection problem. *Archives of Computational Methods in Engineering, 28*, 1103–1127.
3. Nick, A., & Lee, G..(2017). *Cloud computing principles, systems and applications*. ISBN: 978-3-319-54645-2 (pp. 1–378).
4. https://datatracker.ietf.org/doc/html/rfc7539
5. https://nevonprojects.com/secure-file-storage-on-cloud-using-hybrid-cryptography/
6. Reece, B. D., & Ruby, D. (2020). Secure file storage on cloud using hybrid cryptography. *International Journal of Engineering and Technical Research, 8*(03).
7. Jena, A., Das, A. K., Mohapatra, H., & Prasad, D. (2020). *Automated software testing foundations, applications and challenges*, ISBN:978-981-15-2455-4 (pp. 1–165).
8. Ellims, M., & Jackson, K., *ISO 9001: Making the right mistakes*. SAE technical paper series 2000-01-0714.
9. Davis, M., & Weyuker, E. J. (1998). Metric-based test data adequacy criteria. *Computer Journal, 13*(1), 17–24.
10. Jorgensen, P. (1995). *Software testing a craftsman's approach*. CRC Press.
11. Lyu, M. R., Huang, Z., Sze, S. K., & Cai, X. (2003). An empirical study on testing and fault tolerance for software reliability engineering. *Proceedings – International Symposium on Software Reliability Engineering, 2003*, 119–130.
12. Cabe, M., & T. J. (1976). A complexity measure. *IEEE Transactions on Software Engineering, 2*(4), 202–213.
13. Richard, D. J., & Thompson, M. C. (1993). An analysis of test data selection criteria using the relay model of fault detection. *IEEE Transactions on Software Engineering, 19*(6), 533–553.
14. Torkar, R., Mankefors, S., Hansson, K., & Jonsson, A. (2003). An exploratory study of component reliability using unit testing. In *Proceeding 14 international symposium on system reliability engineering* (pp. 227–233).
15. Zhu, H., Hall, P. A. V., & May, J. H. R. (1997). Software unit test coverage and adequacy. *ACM Computing Surveys, 29*(4), 366–427.
16. Voas, J. M., & Miller, K. W. (1995). Software testability: The new verification. *IEEE Software, 12*, 17–18.
17. Zhu, H. (1996). A formal analysis of subsume relation between software testing adequacy criteria. *IEEE Transactions on Software Engineering, 22*(4), 248–255.

Index

M. Khari et al. (eds.), *Optimization of Automated Software Testing Using Meta-Heuristic Techniques*, EAI/Springer Innovations in Communication and Computing, https://doi.org/10.1007/978-3-031-07297-0

The manufacturer's authorised representative in the EU is Springer Nature Customer Service Centre GmbH, Europaplatz 3, 69115 Heidelberg, Germany. If you have any concerns regarding our products, please contact ProductSafety@springernature.com

Printed and bound by CPI Group (UK) Ltd, Croydon, CR0 4YY

15/07/2026

02167632-0001